The Effective Scientist
A Handy Guide to a Successful Academi...

What is an effective scientist? One who is successful by quantifiable standards, with many publications, citations, and students supervised? Yes, but there is much more. Truly effective scientists need to have influence beyond academia, usefully applying and marketing their research to non-scientists.

This book therefore takes an all-encompassing approach to improving the scientist's career. It begins by focusing on writing and publishing – a scientist's most important weapon in the academic arsenal. Part II covers the numerical and financial aspects of being an effective scientist, and Part III focuses on running a lab effectively. The book concludes discussing the more entertaining and philosophical aspects of being an effective scientist.

Little of this material is taught in university, but developing these skills is vital to maximise the chance of being effective. Written by a scientist for scientists, this practical and entertaining book is a must-read for every early career-scientist, regardless of speciality.

COREY J. A. BRADSHAW is the Matthew Flinders Fellow in Global Ecology at Flinders University in Adelaide, South Australia. He has published over 260 peer-reviewed scientific articles, nine book chapters and three books, in addition to his research being regularly featured in Australian and international media.

The Effective Scientist

A Handy Guide to a Successful Academic Career

COREY J. A. BRADSHAW
Flinders University of South Australia

With illustrations by
RENÉ CAMPBELL
Flinders University of South Australia

CAMBRIDGE
UNIVERSITY PRESS

University Printing House, Cambridge CB2 8BS, United Kingdom

One Liberty Plaza, 20th Floor, New York, NY 10006, USA

477 Williamstown Road, Port Melbourne, VIC 3207, Australia

314–321, 3rd Floor, Plot 3, Splendor Forum, Jasola District Centre,
New Delhi - 110025, India

79 Anson Road, #06-04/06, Singapore 079906

Cambridge University Press is part of the University of Cambridge.

It furthers the University's mission by disseminating knowledge in the pursuit of
education, learning, and research at the highest international levels of excellence.

www.cambridge.org
Information on this title: www.cambridge.org/9781107171473
DOI: 10.1017/9781316779521

© Corey J. A. Bradshaw 2018

This publication is in copyright. Subject to statutory exception
and to the provisions of relevant collective licensing agreements,
no reproduction of any part may take place without the written
permission of Cambridge University Press.

First published 2018

Printed in the United Kingdom by TJ International Ltd. Padstow Cornwall

A catalogue record for this publication is available from the British Library

ISBN 978-1-107-17147-3 Hardback
ISBN 978-1-316-62085-4 Paperback

Cambridge University Press has no responsibility for the persistence or accuracy
of URLs for external or third-party internet websites referred to in this publication
and does not guarantee that any content on such websites is, or will remain,
accurate or appropriate.

Contents

Preface

I know the last thing that a young or up-and-coming scientist wants is to have to read *yet another* book, for there are far too many things one must read these days to stay on top of your science. Keeping up to date with the latest articles is difficult enough – if not impossible – let alone reading all that peripheral literature dealing with how to improve your writing, increase your mathematical skills, and communicate with the public. As for reading for pleasure, I will wager that a fair few of you have not read a good novel in years, or you have been stuck on the same one for months. More is the pity of the modern information overload.

So why would you want to read this book? Having read even up to here you have *ipso facto* either already bought the book and intend to read it, or you are currently contemplating the purchase while standing in the aisle of your local bookshop (or hovering over the 'purchase' button on your favourite book-seller's website) and sipping a coffee from the café around the corner. You are busy, your coffee is almost gone, and unless I get to the point soon, you will put down this book and find something more entertaining to read.

So, I will get right to the point – if you are a young scientist, or are planning to become one someday, you will absolutely need to know quite a few of the things that I describe in this book if you want to be an *effective* scientist. What do I mean by *effective*? While I attempt a subjective definition of that adjective in more detail in Chapter 1, suffice it to say here that *effective* is slightly more complex a concept than *successful* or *good*. I argue that all the scientific knowledge, skill, and insight in the world will not be sufficient to ensure that you will indeed be an *effective* scientist unless you master these other issues as well.

The irony is that practically none of this subject material will ever be taught to you formally while a student at university, or even during your first few years as a full-fledged (i.e., employed) researcher. It might seem a wee bit crazy that these essential scientist skills are not taught to you directly, and I most certainly agree with your shocked assessment. Let us not contemplate the 'why not' right now though, for the inadequacy of the modern tertiary educational system would require many more books to cover sufficiently. It is enough to say that you will need these skills during your career development, and it is probably best that you start to acquire them now.

That is why, in hindsight, I wish that I could have read this book, had it existed, when I was a young scientist, and it is the main reason I have now written it myself. Another few questions forming in your mind at this very moment are how I came to acquire these skills, and what gives me the right to claim that I can teach them to you? I have not been awarded a Nobel Prize, even putting aside for the moment that it is not even possible in my chosen discipline (environmental science). Let me be honest – even if there were a Nobel Prize category for my branch of science, I probably never would win it. Most of you reading this book will probably never win one either, so in that sense I am perfectly qualified to lecture you. I have, however, been reasonably successful in the following categories of academic expectation:

I have published a few hundred articles in peer-reviewed journals.
I have been awarded many tens of millions of dollars in research grants.[1]
I have been the head of a successful lab group for many years.
I have supervised many doctoral and masters students.
I have hired and supervised many postdoctoral fellows.
I have spent a lot of time talking to journalists and having my work featured in newspaper, radio, and television media.
I am an avid user of social media to communicate my research to a larger slice of the general public.
I have given many hundreds of public and private lectures.

[1] Most often in conjunction with a team of collaborators.

I have attended and spoken at several scores of conferences, congresses, colloquia, and workshops.

I have won several academic prizes and awards.

I have engaged government and industry partners to influence policy.

In short, and far from boasting (although I suspect many of you will think that I am), I am merely attempting to demonstrate that I have done most of the things that have been *expected* of me as an academic scientist. If I had *not* done all of these things for someone at my stage of his career, then you might even claim that I had not been *effective*. My point here is that knowing what or how to achieve what you will be expected to do is not necessarily intuitive or easy.

I am not going to tell you how to do the actual science in your chosen discipline effectively, for how could I proffer advice for anything outside of my own chosen speciality? Instead, I will outline in the most engaging way I know the lessons I have learned over the years about what leads to good outcomes for academic science in the categories I mentioned earlier.

So, if you are on a career path towards academic science, then I guarantee that you will find this book useful. My speciality is also largely irrelevant here; I think that most of the components in this book apply more or less to most other scientific disciplines, albeit their relative importance might vary to a degree. I will also claim that even if you are not destined for academia, be it as an industry, government, or consultant scientist, there are still many tricks I will divulge that you will probably find useful. It would be impossible to write a book that would appeal to every scientist equally, but I will try my best to keep things as generic as possible while drawing on examples from my own and my colleagues' experiences.

I suppose the 'ideal' reader is someone who is just starting to enter the world of academic research in science, be that a masters or doctoral student, or even a postdoctoral fellow. But if you are not yet or no longer in any of those stages, please do not put down this book just yet. Even though I did not know it at the time I was an undergraduate student, my eventual path towards career academia would have been greatly assisted by the knowledge contained in these pages, and I desperately wish it had not taken me so long to figure out many of the secrets herein when I was a young academic employed for the first time at a university. It is also possible that maybe, just maybe, even the fine, older wines amongst you will still take home a few pointers from this book.

It is my sincerest wish that you have a prosperous, successful, and *effective* academic career.

Acknowledgements

I extend my heartfelt thanks to my employer, Flinders University, for financial support and encouragement to complete this book. I am also grateful to Alan Crowden for encouraging me to write the book in the first place, and essentially crafting the proposal to Cambridge University Press for me. My editors at Cambridge University Press, Lindsey Tate and Dominic Lewis, have given me excellent guidance and the freedom to write the book as I saw fit. I thank Paul Ehrlich, Barry Brook, Franck Courchamp, Irene Driscoll, William Laurance, Nico de Bruyn, Alejandro Frid, Salvador Herrando-Pérez, Paul Willis, and all of my previous students and postdoctoral fellows for many of the ideas presented herein, and I am beholden to the community of followers of my blog, *ConservationBytes.com*. Mar Cabeza, Tomas Roslin, Enrico Di Minin, Sarah Bray, Ben Lewis, Freddie-Jeanne Richard, Deepa Agashe, Lloyd Davis, and Nick Dulvy have all given me opportunities to present elements of this book to students and early-career scientists around the world – with their feedback, the content is much richer than it would have been otherwise. My wife, who has helped me through many of the more personal hurdles associated with academic life, deserves special recognition, as does my happy daughter for her relentless positivity. Finally, I owe a great debt to René Campbell, the highly talented artist and scientist whose illustrations have brought this book to life.

I   What Is an 'Effective' Scientist?

The more I have tried to answer this question, the more it has eluded me. Before I even venture an attempt, it is necessary to distinguish the more esoteric term 'effective' from the more pedestrian term 'success'. Even 'success' can be defined and quantified in many different ways. Is the most successful scientist the one who publishes the most papers, gains the most citations, earns the most grant money, gives the most keynote addresses, lectures the most undergraduate students, supervises the most PhD students, and appears on the most television shows, or is it the one whose results improves the most lives? The unfortunate and wholly unsatisfying answer to each of those components is 'yes', but neither is the answer restricted to the superlative of any one of those. What I mean here is that you need to do reasonably well (i.e., relative to your peers, at any rate) in most of these things if you want to be considered 'successful'. The relative contribution of your performance in these components will vary from person to person, and from discipline to discipline, but most undeniably 'successful' scientists do well in many or most of these areas.

Most of these success measures are easily quantifiable: we can sum up the number of papers you publish, citations they acquire, grant monies you win, students you lecture and supervise, presentations you give, and times you appear on television and are mentioned in the newspapers. But the scientist who spends more time in front of the television camera to the detriment of publishing articles in peer-reviewed scientific journals would not be considered 'successful' by her scientist peers, nor would a highly published scientist be considered 'successful' if none of those articles was cited. So, dominance in one category that comprises success ends up being a curse elsewhere, just as mediocre performance in all these aspects will also compromise at least the appearance of success.

So, being successful in science is achievable if you work hard, follow some basic guidelines, and have at least a modicum of organisational skill (the latter is possibly the most elusive for many scientists I know, including me). In fact, many of the following chapters of this book will describe ways to maximise your potential success. To me, being 'effective' means that you must not only be successful, but that you must also do more than what is normally considered part of the scientist's remit. If your academic measures of success do not extend beyond academia, one could argue that your scientific contributions to society are minimal. This distinction therefore implies that effectiveness has a utilitarian aspect beyond mere academic accolades, but it encompasses a wide array of applications. I am certainly not suggesting that only the classically 'applied' sciences allow the derived knowledge or facts to be 'effective' (Chapter 23), for even purely theoretical disciplines can have huge benefits to society (even if they are not necessarily obvious at the time). Moreover, many of these societal benefits cannot be quantified easily, for one usually cannot measure the benefits of a good education (e.g., the capacity to think logically, to make more evidence-based decisions in life, to be more open-minded, etc.), so the essential role of direct lecturing and mentoring the next generation of thinkers is beyond doubt.

But merely supervising PhD students or lecturing undergraduates does not guarantee a societal benefit, for you might end up producing nothing more than other people capable of performing within the expectations of academia. To go beyond academia is not simple; to catch the attention of people outside of your particular scientific speciality – let alone non-scientists (i.e., the general public) – is an enormous challenge even if the scientific results are exciting to you. We must compete in an age of on-tap information, most of which is either meaningless drivel or utter nonsense. To *market* your scientific discoveries is something for which most scientists are not trained, and so a good deal of this book aims to assist you in this endeavour.

But even marketing is insufficient if your ultimate aim is to improve society, so useful application remains the greatest challenge for any scientist. Irrespective of your chosen scientific discipline, I hope to give you a few additional tools to improve the odds that your science extends past the tip of the academic nose.

This book therefore takes an all-encompassing approach to improving the scientist's career, divided into five thematic parts for a total of 25 chapters. Part I focusses on writing and publishing – a scientist's most important weapon in the academic arsenal. I start with advice on how to be more than just a writer of mundane scientific prose (Chapter 2), to tips on good writing practices (Chapter 3), and article-writing strategies (Chapter 4). From there we head into the murky realm of co-authorship (Chapter 5), and how to choose the best outlets to maximise your publication track record (Chapter 6). Next is a chapter discussing strategies for dealing with rejection and peer criticism (Chapter 7), and I round off the part by delving into the delicate art of peer review and constructive editorship (Chapters 8–9).

Part II deals with the numerical aspects of being an effective scientist, from good analytical skills (Chapter 10), to efficient data management (Chapter 11) in this age of exponentially increasing data generation and availability, and finally, how to obtain and manage your research monies (Chapter 12). Part III is focussed on running an efficient lab, which includes the politics of good lab relationships (Chapter 13), the quality of mentoring students and staff (Chapter 14), actively promoting gender equality and your lab's cultural diversity (Chapter 15), suggestions for efficient time management (Chapter 16) and practicable work–life balance (Chapter 17), and finally, how to reduce the inevitable stress that will threaten your mental health by following all the recommendations in the previous chapters (Chapter 18).

Part IV covers the more entertaining aspects of being an effective scientist, from the best practices for making scientific presentations that people will stay awake for and remember (Chapter 19), making the most of scientific conferences (Chapter 20), getting your scientific messages out to the rest of the (non-scientist) world (Chapter 21), and tips for dealing with the press (Chapter 22). Part V raises some more philosophical aspects of a scientist's moral duties and societal role, including an exposé on making your science as useful as possible to society (Chapter 23), the touchy subject of science and advocacy (Chapter 24), and some final thoughts on what being a scientist is all about anyway (Chapter 25).

PART I Writing and Publishing

2   **Become a Great Writer**

Meeting someone for the first time at whatever social event I happen to be attending and where polite conversation ensues, I am inevitably asked this question 'What do you do (for a living)?' I usually just respond 'scientist' and gauge whether my new companion shows even the slightest sign of desiring further clarification (most do not), or would rather look over my shoulder at the drinks table or at someone more aesthetically pleasing than I am (most do). But if they do happen to grace me with even a feigned interest to elaborate, I end up going down one of several explicative pathways depending on my mood. To some, I am an 'environmental scientist', to others, an 'ecologist', and to a few I claim to be a 'mathematician'[1] (more on that later). However, if I am honest about how I spend most of my professional life, and a disconcertingly large part of my personal time, it would be more accurate to say that I am a *writer*.

I suspect that most people might imagine a writer primarily as someone who pens novels, or at least who is consulted to write articles for various magazines, newspapers, or online media. But I wager few would have 'scientist' immediately come to mind as a type of writer. Of course, scientists often do the more stereotypical tasks of looking through microscopes, designing and tending to laboratory experiments, collecting data in the field, writing on clipboards or tablets, and analysing data on a computer. However, if that is *all* that they do, I argue that they are not scientists per se; rather, they are technicians, assistants, or research officers.[2] To be a true scientist, you

---

[1] Although to a real mathematician, this could be considered slightly insulting.
[2] There is nothing wrong with this, of course; such people are extremely important.

have to spend a good deal of your time putting all of those data, analyses, and results into legible sentences that clearly explain what you have done, what you have discovered, and what they probably mean. Those words are then scrutinised by other scientists, and then they are published in – most often – scientific journals. The truth is that if you are an academic scientist, this process of writing will probably take up most of your time.

So, scientists are writers, and it goes without needing extensive justification that good writers must have good writing skills. This of course means *inter alia* possessing an expansive vocabulary, spelling correctly, and understanding the subtleties and complexities of good grammar, but it also means presenting complex concepts clearly, defining terms precisely, avoiding jargon, and generally being able to prevent the reader from falling asleep after the first few paragraphs of your introduction. Even though scientists are *de facto* writers, most fail miserably at most or all of these aspects, which is why most scientific writing is exceedingly dull and uninteresting to all but a handful of specialists.

If you accept this logic, to become an effective scientist you therefore have no choice but to master the art of good writing. These days nearly all major scientific writing is done in English, so even if you write well in Chinese, Arabic, or Spanish, you have the additional burden of learning how to do so in English. As it turns out, non-native English speakers are not necessarily at a disadvantage here because perfecting another language is universally good for the brain (1, 2), and actively learning the nuances of English grammar as an adult often leads to a better grasp of the written language than even most native speakers inherently possess. In fact, there is little support for the hypothesis that not being a native English speaker weakens your long-term scientific performance (3).

Most scientists either actively chose, or passively ended up doing, coursework at school and university that, funnily enough, focussed on scientific topics such as biology, chemistry, mathematics, and physics. Although many were required to take at least some language courses, most probably placed much more emphasis on the science subjects at the expense of the language subjects. I am still painfully aware of this oversight in my own education to this day, because as I have made it abundantly clear, I am now principally a writer.

One might assume that the greater scientific community has had plenty of time to figure this out, but instead it seems that there is a waning interest in teaching and maintaining good writing skills. So obvious to me is this trend that I have noticed an unfortunate and steady decline in the capacity of incoming students to write English well. Before the grounding in scientific theory – whatever the discipline – English-writing prowess is an absolute necessity in scientific training. Unfortunately, this skill has been slowly putrefying in the bowels of our educational systems for decades. I admit that I am something of a stickler when it comes to good grammar, and more than once I have been called a pedant for it. Regardless, I maintain that without good, precise English, the incontrovertible communication of one's hard-won scientific data is often lost in the lazy mists of poor grammar, archaic turns of phrase, terminological imprecision, and bad writing habits.

As a university professor, I am convinced that I can teach nearly anyone the basics of my chosen scientific field, and even get across the most complex theories to the uninitiated. But to teach someone to write! Are there enough hours in the day or days in the year? One could also question whether this is indeed my job – am I now responsible for the instruction of basic grammar? Increasingly, it appears that I am.

I will cite many examples of this problem from Australia because that is where I have spent most of my professional life, but many apply as well to other countries. Ask almost any Australian under the age of thirty if he or she can identify a noun, adjective, or adverb (let alone a dangling participle, indefinite article, or split infinitive) in any given sentence, and I will wager a large sum that most cannot. But how important is it to be able to identify by name the constituents of language? Who should care, as long as one can write well? The simple fact is that none of those ignorant in the ways of grammatical nomenclature can write any better. This is because in places such as Australia, grammar has not been taught properly in most schools since the late 1960s or early 1970s; how can one have a functioning and enduring society if few have the tools to be able to write about their ideas clearly? I liken the problem to the example of an architect versus a carpenter. Ask the architect to build you a house – while it might look good on the drafting paper, the first strong breeze would probably knock the building over. Instead, if you ask the carpenter to do the job, the doors will probably fit and the floor will support the weight of more than one human being, but chances are that the edifice will not make him a contender for the Pritzker Architecture Prize. Only the one that knows how to use the tools properly for the job will persevere.

Do not despair if you suspect that your writing skills are inadequate; but for the love of all things beautiful in language and for the sake of unambiguous scientific interpretation, do something to rectify this. If you know that you are not strong in English, the onus is on you to fix the problem. Take courses, read more books on the basics of language (of which there are many fine examples) (4, 5), and practice writing both prose and poetry – for these skills will continue to become more and more relevant as your career progresses, and you will require better and better flourish of your virtual pen (keyboard) as you develop into a mature scientist. I contend that English, even before mathematics, is the cornerstone of all good science. Without it, you will be mediocre at best, and unintelligible at worst.

TERMINOLOGICAL CHAOS

But science writing is not poetry; while concepts and evidence are often open to interpretation, scientists must strive above all things to be precise in their meaning. Although science is fundamentally mathematical (Chapter 10), we generally do not speak or write entirely in terms of equations.[3] Mathematics is perhaps one of the most precise of human inventions, because $1 + 1$ always equals 2, but many words do not have the same degree of terminological precision. Therefore, when scientists use words in a style subject to few rules (apart from grammatical), the odds of distorting the scientific meaning are high. The problem is compounded by the exponentially growing mass of scientific articles appearing in all disciplines, meaning that while total human knowledge is growing, ironically, individual scientists know proportionally less of their chosen field with each passing year. There are simply too few hours a day to keep abreast of all that new information.

The outcome is that scientists tend to re-invent terms with astonishing frequency, such that there are now many subtly different terms to describe the same essential concepts, no matter what the discipline. For example, in one sub-field of my particular area, we found over forty terms for four discrete concepts (6), which represents an inflation of jargon arising because even within the same discipline so few scientists are familiar with their peers' work. In fact, there are four main scenarios causing the proliferation of jargon:

Synonymy
  'Lift' and 'elevator' are the same thing (depending on where you grew up in the English-speaking world, or who taught you the language). Different authors have given different names to the same established relationships, because they believe their terms are better than existing ones, or because they are unaware that the concept's terms had already been coined elsewhere. This situation partly reflects the problem that many scientists work secluded in their specialised areas of expertise.

Polysemy
  'Asteroid' also means 'comet'. Here, researchers re-invent definitions of existing terms, indicating that they have not read or taken into consideration the foundational literature where the terms were originally coined and defined. This problem fits a general trend of

---

[3] Although the casual perusal of many scientific articles published in mathematical or statistical journals might suggest otherwise.

neglect of older literature in which many hypotheses and theories of our modern science were laid out many decades ago.

Inflation

Modifying the letters in 'molecule' to spell 'culemole' or 'mellocue' does not equate to the original meaning. Some scientists, to explain their personal views regarding a debated phenomenon, have created a battery of new terms with a strong philosophical flavour. Such controversies, and the resulting jargon have impeded rather than progressed many disciplines.

Refinement

'Female' and 'male' differentiate types of 'human being'. As we progress in our understanding of how nature works, new concepts and new terms to name them are genuinely needed.

This messy quadruplet (synonymy, polysemy, inflation, and refinement) is now a dominant feature of much of the scientific literature. This becomes apparent when experts publish reviews of the history, evolution, and involution of particular terms, or when attending a conference where two speakers (apparently) disagree or talk about different topics because they are using a different lexicon. But terminology through intellectual dispute reflects, not concepts, but schools of thought; yet, terminology driven by opinion rather than empirical or theoretical facts contributes little to scientific progress.

Poor terminological standards are the unseen bane of students and early-career researchers trying to understand and apply concepts on which the very experts disagree – the legacy of individual authors also relies on how understandable their published work is to the younger (but future) generations of scientists. Certainly, the understanding and expression of scientific information by non-English speakers is seriously jeopardised by inconsistent terms – even the translation of specialised treatises or textbooks can be problematic because translation relies on context rather than terms – so how do we translate a term created in English into Spanish or Arabic, or *vice versa*?

So how do we avoid this terminological chaos? Can any one scientist do anything to limit the confusion? The answer is both 'no' and 'yes', because although it will be increasingly difficult to keep up-to-date of your field's developments, a focus on terminological precision can certainly be encouraged and embraced. At the very least, a term should be adequately researched within the historical literature to understand what components it includes and excludes. Assuming a particular meaning can be dangerous unless you have read the term's source. This means that when writing your own article, you should provide a clear, well-referenced definition of your chosen terms to avoid inadvertently ascribing alternate or variable meanings. Some journals also have the option of including a glossary of terms with the article, which is an opportunity that you should never miss. Glossaries and clear definitions notwithstanding, avoid the temptation to invent new terms because chances are, someone has already done it before you.

Some scientific disciplines, such as astronomy, biochemistry, medicine, and taxonomy, have even taken the bold step of regulating their terminology, because unique names for a star, molecule, heart disease, or species are critical for application, sharing, and review of existing knowledge, and for avoiding redundant research (7). Terminology merits due recognition because it is the basis of the classification of knowledge, and so indicates the state of progress and maturation of any science. If you are lucky enough to work in a scientific discipline with just such a terminological convention, make certain you know and understand how terms are applied. If you are not, I strongly recommend that you promote the development of terminological conventions within your field, else the problem will only get worse.

BREAKING BAD HABITS

Although there are far more comprehensive guides to improving your writing in a general sense, in this section I provide a list of common errors, unnecessary jargon, bad phrasing, archaic usage, and overly complex constructions that I often see in scientific writing. Many of my suggestions are personal preferences, but I try to justify them in each case. Some issues apply to English writing in general, others to science only, and others just to certain fields. In most cases, these problems are merely bad habits that have been passed down the scientific generations either actively by academic supervisors to their students, or passively by many years of reading poorly written scientific articles. Change happens slowly, but if you can nip some of these problems in the bud before they become habits, you will be more likely to write better scientific articles that more people will read and, ultimately, cite in their own work.

a number of

This is just meaningless, sloppy writing. If you write '...a number of researchers contend that...' instead of '...many researchers contend that...', you are just wasting words. Taken literally, '...a number of...' could also mean 'zero', because 0 is also *a number*. Better yet, just quantify the items in question and avoid the ambiguity.

abbreviations, acronyms, and initialisms

An *abbreviation* is a shortened version of a full word (e.g., Prof. [Professor] Smith); an *acronym* is a type of abbreviation that can be formed into a new word (e.g., laser [light amplification by stimulated emission of radiation]); an *initialism* is an abbreviation, usually consisting of the first letters of a string of words, but that does not itself create a new word (e.g., IUCN [International Union for the Conservation of Nature] – you do not say 'i-yooo-s-in' when pronouncing it; IPCC [Intergovernmental Panel on Climate Change]). Sometimes, abbreviations can be either acronyms or initialisms depending on personal preference (e.g., CSIRO[4] [Commonwealth Scientific and Industrial Research Organisation], pronounced either as 'C.S.I.R.O.' or 's-eye-r-oh').

Most scientists appear to *love* AAIs (See that? I just created a new initialism), with the unfortunate corollary that much of their writing is impenetrable to all but a handful of similarly trained people. The typical belief is that because AAIs are much shorter than the clunky strings of words they represent, writing is streamlined and much clearer to understand. The reality however is that it does just the

---

[4] Australia's semi-autonomous (from government), national applied-research organisation.

opposite because without an extreme familiarity with the new term, the reader is constantly required to refer to the first time the AAI was defined. This means it cuts the flow of reading and comprehension. The take-home message is therefore to use these structures sparingly, even at the cost of a slightly longer document of text. Only in cases of great familiarity should these be used at all (e.g., NASA [National Aeronautics and Space Administration]). An obvious exception is for units of measurement that are standardised for scientific purposes by the *Système international d'unités* (SI, translated in English as the International System of Units) – e.g., 'm' for 'metre', 'kg' for 'kilogram', 's' for 'second', and 'K' for 'kelvin'. These do not usually need to be defined or spelled out, unless they represent a rarely used or unfamiliar unit of measurement.

cf.

This is an abbreviation of 'confer', meaning 'to compare'. It is one word, so its abbreviation requires a single full stop after the 'f'.

compound adjectives

This is a particularly abused element of scientific writing. Although the rules are straightforward, I am amazed just how many people get it wrong. Most scientists appear to understand that when an adjective (that is, a qualifier for a noun, just in case you are a grammarling) is composed of more than one word, there is normally a hyphen that connects them:

e.g., '10-m fence', 'high-ranking journal', 'population-level metric', 'cost-effective policy'

If two or more adjectives are given in a row, but none modifies the meaning of the others, then it is simply a case of separating them with commas:

e.g., 'a long, high fence', 'an old, respected journal', 'an effective, enduring method'

However, if the compound adjective is composed of a leading adverb (that is, a qualifier for a verb), then there is NO hyphenation:

e.g., 'an extremely long fence', 'a closely associated phenomenon', 'a legally mandated policy'

There are other instances when no hyphenation is required, such as when the qualifiers are proper nouns (e.g., 'a Shark Bay jetty'), from another language such as Latin (e.g., an '*ab initio* course'), or enclosed in

quotation marks (e.g., 'a "do it yourself" guide'). Note in the last
example, without the quotations, it would become 'a do-it-yourself
guide').

A quick way to recognise whether a compound adjective should be
hyphenated is to examine the terminal letters of the leading word; if
the leading component ends in 'ly', then it is likely (but not
necessarily) an adverb, and so the compound should not be
hyphenated.

contractions

Words such as *can't, won't,* and *it's* are colloquial forms and should never
be used in a scientific manuscript.

conduct

As in '...we conducted the experiment...'. There is absolutely nothing
wrong with the simpler 'do' or 'did'. I have never seen a scientist
'conduct' anything, but I have seen a few good orchestral
performances where a person whose back is turned to the audience
keeps all those talented musicians in line. Do not blindly fall for the
false presumption that fancy, less commonly used words are in any
way more 'scientific' than simple, commonly used words.

critical(ly)

As in '...highlights the critical importance of...', this term is generally
meant to communicate some urgent need or necessity. While most
authors would like to think their chosen topic is 'critical', many
neither define to whom or to what the results are 'critical', or even
what the lack thereof would mean. In some circumstances, it is used
to infer some sort of threshold beyond which another state dominates.
If you are trying to inflate the importance of your work, 'critical' is the
word to use; if you mean a 'threshold', then just use the latter term.

data

I suspect that I might be losing this battle because the incorrect
conjugation of verbs following this word has become so ubiquitous
that I suspect scientists are now also beyond rescue. Nonetheless, the
word 'data' should always be followed by the plural conjugation of
verbs (e.g., '...the data are...'; '...the data were...'). A singular
'datum' is one measurement and requires the singular conjugation. A
'dataset' (or 'data set') is a single group of data, so it too can use the
singular conjugation. If you want to communicate that your sample

size was too small (for your intended purposes), you need to write 'too few data' (see *few* versus *less*).

decadal apostrophes

I do not know how this trend got started, but it appears to be an affliction related to the abuse of apostrophes in general. When referring to a specific decade (e.g., 'the nineties') in numeric format ('1990s'), there is no reason to insert an apostrophe, UNLESS it is meant to be possessive. For example: '...we acquired data from the 1990s to the 2000s to...', or '...many researchers use 1990s' approaches to investigate...' (note that the apostrophe is placed after the terminal 's' because we cannot write *1990s's* in English).

decimate

As in '...the population was decimated following...'. I have seen this one used way too often. It is usually invoked by the author to imply some devastating reduction in size (somehow it sounds bad); for this reason alone, the emotive language should be avoided. However, 'decimate' has a specific meaning: to reduce by every 'one in 10' (hence the *deci* prefix). If you really mean that something was reduced by 10 per cent, use 'decimate'. If you are just stating that something was reduced, state by how much and avoid emotive and incorrect terms.

dramatic(ally)

As in '...we observed a dramatic decline in...' or '...the concentration increased dramatically...'. This is another meaningless, emotive word that belongs to the theatre, not in scientific writing. Quantify your meaning instead of relying on subjective terms.

few versus less

I am flabbergasted that this still stumps so many people. 'Few' should be used to define a small number of countable (discrete) items (e.g., individuals, quadrats, plots, experiments). 'Less' should be applied to a measurable, continuous variable (covariate) that cannot be easily discretised (e.g., water, biomass, carbon, electricity). If you ever see someone write 'less individuals', take out the big red pen.

has been shown to

As in '...is a phenomenon that has been shown to demonstrate a...'. There is no need for this verbiage. Simply state '...is a phenomenon that demonstrates a...' and then reference the statement properly at the end of the sentence.

i.e. and *e.g.*

These abbreviations of the Latin, *id est* and *exempli gratia*, literally mean 'that is' and 'for the sake of example', respectively. They are two words, each abbreviated, so a full stop is required after each letter in the abbreviation. Absolute correctness normally dictates the addition of a comma after the final full stop, but many journals drop the comma for whatever reason.

in order to

As in '...in order to compare the plots...'. What's wrong with just 'to'? I have rarely seen a situation requiring 'in order to', so usually it represents unnecessary verbiage. However, 'in order to' can be used sparingly if the first action is done particularly for a specific intention, and this idea needs emphasis.

its and *it is*

The answer to why it is so difficult for people to understand this one has forever eluded me. In almost every other circumstance, an apostrophe followed by an 's' indicates possession to a singular noun, as in '...the transect's divisions...', '...the nearest neighbour's value...', etc. When the noun in question is plural, the apostrophe sits nicely outside the terminal 's' (e.g., '...the species' attributes...'). This is a quasi-universal law EXCEPT for *its*/*it is*. In this case *it's* is the contraction of *it is*, so *its* becomes the possessive form. So, you can write '...its trend...', but '...it is trend...' is clearly incorrect. Still confused? There is a simple way to remember – whenever you see *it's* in front of something, say *it is* to yourself and see if the phrase makes sense. If it does not, then it should be *its*.

key

Used as an adjective (e.g., '...a key determinant of the strength of feedback is...'), it generally fails the definition of 'key'. A 'key' is something that unlocks something else. The weaker case of using nouns as adjectives notwithstanding, writing 'key' as an adjective should only be done when it describes something essential for a phenomenon to exist. Writing 'key' just to make it sound more important is unnecessary sensationalisation.

level

A 'level' is a discrete value; therefore, it cannot be applied to a continuous variable. So, writing 'carbon dioxide levels' or 'high levels of contaminants' is incorrect. Statistically speaking, a 'level' is a discrete value of an ordinal factor (e.g., 'medium' is a discrete value of a three-level factor made up of 'small', 'medium', and 'large').

majority

This is a similar problem to 'a number of'. If you mean 'most', then write 'most'. Better yet, quantify instead of referring vaguely to a value greater than 50 per cent.

may versus **might** or **can**

I have often written this incorrectly too. 'May' implies doubt or permission, so it is most often better to use 'can' or 'might' (where appropriate) when you expressly mean 'under certain circumstances'. If you ask yourself whether the object of the sentence should be consulted and can choose what to do, then 'may' applies (e.g., '...some researchers may use method X...'. If the object does not really have a choice, then 'might' or 'can' should be used (e.g., '...the star might explode...').

myself

Using 'myself' instead of 'me' or 'I' has become something of a grammatical plague lately; I hear or see it almost every day in radio interviews, newspaper quotes, and e-mails. I think it originates from that peculiar trend a few decades ago where people who did not understand the 'I' rule overcorrected so much that 'me' became the embodiment of grammatical evil itself:

'...my colleagues and I wrote a paper that...' (here, 'I' is the subject; just remove the '...my colleagues and...' to see if it makes sense)

'...the paper written by my colleagues and me...' (here, 'me' is the object; again, remove '...my colleagues and...' to see if it makes sense).

However, today many people confused by this terribly simple rule avoid the problem altogether by inserting 'myself' wherever 'I' or 'me' is warranted. I have even had people write to me in e-mails '...just send it to myself...'. Whatever did 'me' do to be vilified so?

perform

While some scientists might also be accomplished Thespians, it is wise to keep the theatre out of science. See *conduct*.

relatively

In principle, there is nothing wrong with this word, but it is often added to invoke some sort of importance without any point of comparison. 'Relatively' can only be used when it is actually 'relative' (i.e., compared) to something else. You cannot just write '...the phenomenon is relatively important in...' unless you can specify its importance *relative* to something else.

significant

This is a big issue, so I will devote more discussion to it than all the other bad science-writing habits I mention. Most science writing has become burdened with archaic language that perhaps at one time meant something, but now given the ubiquity of certain terms in most walks of life and their subsequent misapplication, many terms no longer have a precise meaning. Given that good scientific writing must ideally strive to employ the language of precision, transparency, and simplicity, now-useless terminology should be completely expunged from our vocabulary. The word *significant* (and her sister noun *significance*) is just such a term.

Most interviews on radio or television, most lectures by politicians or business leaders, and nearly all presentations by academics at meetings of learned societies invoke 'significant' merely to add emphasis to the discourse. Usually it involves some sort of comparison –...a *significant* decline, a *significant* change, or a *significant* number relative to some other number in the past or in some other place, and so on. Rarely is the word quantified: how much has the trend declined, how much did it change, and how many is that 'number'? What is 'significant' to a mouse is rather unimportant to an elephant, so most uses are as entirely subjective qualifiers employed to add an unspecified emphasis to the phenomenon under discussion. To most, *significant* just sounds more authoritative, educated, and erudite than *a lot* or *big*. This is, of course, complete rubbish because some people use big words to hide the fact that they are not quite as clever as they think they are.

While I could occasionally forgive non-scientists for failing to quantify their use of *significant* because they have not necessarily been trained to do so, I criticise scientists who use the word in this way. As scientists, we are specifically taught to quantify natural phenomena, so throwing 'significant' around without a clear quantification (e.g., it changed by x amount; it declined by 50 per cent in two years, etc.) runs counter to the very essence of the scientific method. To make matters worse, one often hears a vocal emphasis placed on the word when uttered, along with a patronising hand gesture, to make that subjectivity even more obvious and unpalatable.

The only possible exception is when using the word in a statistical sense, although depending on your own statistical philosophy, it can be argued that even invoking this meaning of *significant* is no longer useful or meaningful (8). Unless you are a rather young scientist who has had the rare privilege of avoiding 'classical' statistics training, then you are most certainly aware that *significant* is the term used to describe the probability that a particular phenomenon under investigation could have arisen by random chance.

So-called 'significance' tests are known more formally as Neyman–Pearson (null) Hypothesis tests, where a single 'null' hypothesis is 'rejected' based on an arbitrary threshold probability of observing a value of the metric of choice as extreme as the one observed. This probability – the well-known 'p-value' – simply refers to the probability of making a mistake when rejecting the null hypothesis (also known as a 'Type I error'). Most disciplines still cling doggedly to a threshold probability of 0.05 (1 in 20 chance) below which the null hypothesis can be rejected.

If you are an indoctrinated scientist, then this might sound perfectly reasonable, but I wager that it might strike non-scientists as being rather silly. Indeed, it is silly. There is nothing important at all about a 1 in 20 chance of making a mistake. Nothing special happens when your probability of making a Type I error slips below 0.05. In other words, it is absolutely meaningless. Would you cross a busy road if the probability of dying before reaching the other side was 1 in 20? Unless you were suicidal, I doubt it. Would you cross if it was 1 in 21? 1 in 25?

From where did this entrenched, silly number come? As it turns out, it is an unfortunate accident of printing convenience that we use the 0.05 threshold today. It is in fact a holdover from the days when statistical tables had to be printed in the back of textbooks. There was traditionally insufficient space to write all manually calculated rejection probabilities for distribution-specific null-hypothesis tests, so they were often truncated at 0.05 for convenience. It is as simple and forehead-slapping as that.

The last remaining supporters of the Neyman–Pearson Hypothesis Testing paradigm (9) might claim that there is a time and a place for it, as long as we avoid arbitrary binary assessments of 'yes' and 'no', for probability is an infinitely sliced gradient between zero and one. However, the other problem with this paradigm is that Neyman–Pearson approaches cannot simultaneously consider other dimensions of the question, namely, evaluating the relative statistical support for alternative models. Neither can null-hypothesis tests evaluate Type II errors (i.e., the probability of making an error when failing to reject the null hypothesis).

The alternative – and thankfully growing – paradigm is the multiple working-hypotheses approach (10). Instead of considering a single (null) hypothesis and testing whether the data can falsify it in favour of some alternative (which is not directly tested), the multiple working-hypotheses approach does not restrict the number of models that can be simultaneously considered. The approach can specifically accommodate the comparison of hypotheses in systems where it is common to find multiple factors potentially influencing the observations made. This book is not designed to teach you statistical approaches, but I can guide you to several good references to follow up on this issue in particular if you are interested (10–14). Suffice it to say that there are now no more excuses – neither terminological nor statistical – for using 'significant' in scientific writing. So, let us dig a deep grave and bury *significance* permanently. It does no one any good any more.

situated

As in '...our study area was situated in...'. You should simplify this just to 'is' or 'was'.

space between numbers and measurement units

I suspect that this unfortunate fashion originated from professional sign-makers charging their clients by the character (spaces included), which is why you will often see road signs that state 'Rest Area 1km Ahead'. However, English demands that you put a space between '1' and 'km' because you cannot write 'onekilometre' when the number and abbreviation are spelled out. This also applies to all other units of measurement.

split infinitives

I believe the jury is out on the acceptable use of split infinitives, and I confess that this is another grammatical battle I will probably lose. An infinitive (for those of you who are grammatically challenged, an 'infinitive' is the base form of the verb prior to conjugation) can never be split by an adverb in English. How many times have you seen '... to significantly affect ...', '... to adequately measure ...', or '... to properly test ...'. These are all wrong (they should instead be written as '... to affect significantly ...', etc.).

that versus which

I admit that this is not an easy rule to master in English. In the most basic description of the difference, 'that' usually introduces essential information in a restrictive clause, whereas 'which' introduces additional information in a non-restrictive clause.

> 'What is FASCINATING to me is that ... one way to determine ... the correct word ... is to ask the question, "Does the clause clarify which of several possibilities is being referred to?" If the answer is yes, then the correct word to use is that. If the answer is no, the correct word to use is which'. (15)

This might seem somewhat counter-intuitive, but it is correct (hence the confusion).

to the best of our knowledge

These are weasel words that only demonstrate that you have not done your homework. If you have genuinely missed an important reference, your referees will point it out. Do not try to justify your poor research with this nonsense.

utilise

Just write 'use'. For some reason people believe 'utilise' sounds more technical. It does not.

very

> As in '...there are very few occasions...'. 'Very' has no place in scientific writing; I defy anyone to quantify what it means because it has an entirely subjective interpretation. The same applies to the word *quite*.

voice

> If science is the best way to reduce subjectivity when asking a question of how something works, then an inherently essential aspect of this is getting your message across to as many people and as clearly as possible. One of the best ways to write archaically such that few people will even want to read your work is to use the passive voice. Here are a few examples:
>
> > **passive**: The analyses were conducted...; **active**: We analysed...
> > **passive**: The data were collected in accordance...; **active**: We collected the data according to...
>
> Some might think the differences are largely irrelevant, or even that the passive voice sounds somehow more 'science-like' (technical) (16). But I argue that this justification is utter nonsense and exactly WHY you should not use the passive voice. Using meaningless or subjective words like 'significant' or 'conduct' or 'perform' just because they sound more technical is not going to fool anyone, and using the clumsy, archaic, and longer passive-voice form for the same reason is deceptive at worst, annoying, and unnecessary at best. I have even had a collaborator state quite assuredly that 'you will not get published in a British journal if you use the active voice'. What journals would those be, I wondered? I have never had this experience, nor have I ever had a manuscript rejected by using the active voice.
>
> I have never understood why a scientist, who has sweated blood and shed tears to collect, analyse, and present her or his hard-earned data, would then not want to take credit for that effort? Why apply the passive voice to hide your identity? Who are these mysterious scientific automatons who collect and analyse our data for us? 'The data were collected...' – please! Take credit for your work and use the active voice.

I will concede one exception to the near-universal need for active voice – if you use, for example, data that you did not collect, then I suppose it is acceptable from time to time (i.e., infrequently) to use the passive voice. However, one could argue that you can use the active voice as long as you identify the person(s) responsible.

3 Me Time

One of the most basic and enduring recommendations for effective writing is identifying good places and times in which to do it – this advice applies to all forms, themes, and types of writing. In the specific case of academic writing, however, there are a few more nuances to 'finding a quiet place to write' that I will now endeavour to expound.

As in most activities requiring intense concentration, an environment without too many distractions will likely lead to better, more efficient outcomes. Writing in particular requires a singular focus that most people just do not possess innately. I think I am one of those who has always struggled to maintain focus while writing, despite my ability to code mathematical formulae for hours on end – even to the exclusion of relieving myself until near-bursting. I cannot possibly estimate the number of times I have seemingly awoken from a trance-like state after spending hours in front of a computer screen with an intense, penetrating hunger because I forgot to eat my packed lunch. The problem with writing though is that I tend to get distracted too easily, especially in the case of revising manuscripts following peer review. I am convinced that I am genetically predisposed to procrastination. However, with a little practice and by following a few simple guidelines, I can show those of you who are similarly afflicted how to turn these weaknesses into formidable writing tools.

WHEN

We will start with the most obvious criterion – the quiet time. That might seem at least superficially rather straightforward, but when faced with regular requests from your students for office visits, demands by administrators that you must attend yet another useless meeting, the constant 'pinging' of enticing e-mails, tweets, and

Facebook updates (see Chapter 16), and the rumblings of cacophonous herds of undergraduates shambling obliviously through the corridors trying to find that cleverly hidden lecture theatre, it can be downright challenging to get even twenty continuous minutes of uninterrupted writing time. For me, I find that I am utterly unable to write anything of value if I do not have at least one, but preferably two hours of continuous writing time.

Some simple rules for minimising the probability of interruption are intuitive. The first, of course, is to set sensible meeting times if you cannot avoid them entirely. In the great hierarchy of meeting priorities, my highest falls squarely on my postgraduate students, and especially if they want to discuss one of the manuscripts they are preparing for submission. But I do not drop everything immediately at their request – generally I will try to schedule the meeting for first or last thing in the day because it maximises the uninterrupted stretches in between. The same rule of thumb can be applied to almost any meeting for which you have even a little control of timing. Another trick is to schedule meetings on Mondays or Fridays, especially if you are the slow-to-recover-from-the-weekend, or the cannot-wait-until-sundowner-cocktail type, respectively. If those days are largely write-offs for you anyway, you might as well spend them in meetings. Better still is to avoid meetings altogether whenever possible.

For all other times when you are not in a meeting, not lecturing, not responding to or composing e-mail messages, not engaging in social or traditional media, and not taking sustenance, you should ideally schedule blocks of dedicated writing time. You can be as rigorous about this as you like, such as setting particular slots on specific days of the week in your computer's calendar as no-exception writing times. This approach works well for the hyper-organised amongst you, but I am really not that regimented in my own scheduling. I prefer a more organic approach by examining my impending week's schedule and deciding when I could most easily block out specific slots. If you do adopt this sort of scheduling protocol, I further recommend that you stick to it as best as you can, specifically by avoiding to schedule

anything else during these slots. You should not need reminding that
a shut office door attracts fewer unanticipated visitors.

My empirical, cognitive self is now satisfied that I have given
you the sagest time-management advice for writing effectively, but
my organic, limbic self is complaining that I have not paid it enough
attention. In deference, therefore, to that spontaneous spark of apparent genius that can occur seemingly at random, I also strongly urge
you to go with the flow when the word-geysers really start to gush.
If you are anything like me, good ideas often come at the most inopportune and inconvenient times, like when I am in the car, walking
to work, or trying to stabilise my shaky *utthita trikonasana* pose in
yoga class.[1] If I am suddenly struck with the urge to write something
down, and I am in the position to be able to do so, I will try to drop
everything to nourish the whim. Even with the best intentions and no
distractions, sometimes the writing will just not come; at other times,

---

[1]   Yes, I have recently taken up Iyengar yoga as an effective means to manage my stress
(see Chapter 18). Thinking about anything *except* the pose itself is a strict no-no in
Iyengar, but sometimes I cannot help myself.

writing seems as natural as breathing. If you find yourself in a writing groove, do not interrupt yourself for anything until the spirit has passed, for these times are often the most insightful and productive.

THINGS THAT GO 'PING'

It is no exaggeration to claim that I could not do my job as a scientist now without e-mail. Long gone are the days when scientists collaborated mainly because they occupied adjacent offices, or at least worked in the same institution. While I do not yet consider myself to be 'old', I can still remember when we sent manuscripts to our collaborators and to the journals themselves in thick, heavy envelopes stuffed with at least three copies[2] of the double-spaced and single-sided manuscript. The number of trees we destroyed in the quest for scientific advancement was obscene. I also still vividly remember sitting down in front of a communal lab computer and using e-mail for the first time, thinking that this new-fangled communication device was interesting, but largely a waste of time. How I was wrong, although in some ways I was also so very right. E-mail has radically changed the process of scientific collaboration, because I can now distribute my evolving manuscripts to my co-authors around the planet in a matter of seconds, regardless of the size of the files concerned. In fact, I once calculated that I have thus far co-authored papers with more than 400 people whom I have not yet met in the flesh. During nearly every conference I attend, completely unfamiliar people[3] will introduce themselves to me after a quick glance at my name tag with a 'Hi Corey. I'm *so-and-so*. We wrote a paper together'. 'Did we, now?', I think to myself.

So, while e-mail[4] is of critical academic value, it is simultaneously a temporal malediction because it can waste so much of your

---

[2] Sometimes we were required to submit up to seven copies.

[3] Do not waste this opportunity to engage with a potentially 'new' collaborator. See Chapter 20.

[4] And now platforms that are rapidly supplanting e-mail, like Skype, Slack, and Basecamp.

time.[5] The most basic advice I can proffer here is to quit your e-mail program during your allocated writing times, and avoid all temptation to switch it back on until after you have finished with your slot. This applies equally to your internet browser, your Skype/Zoom/Slack/*et cetera* account, your calendar, any other program or app that might try to notify of something, your tablet, your smart phone and of course, your office telephone. In other words, turn off or temporarily disable anything that could siphon your attention away from the task at hand. Remember that you will only be writing for a short time relative to the rest of your working day, so even if you might miss an emergency telephone call or e-mail, it will not be for long. Just relax – the chance of an emergency occurring is remote.

WHERE

Where to write is just as important as *when* and *how*. Most of you will have some type of office in which to do most of your writing, but depending on your career stage, this could mean any sort of sharing arrangement with other scientists who might not necessarily embrace the same writing philosophy. Even if they do, there will inevitably be telephone calls, e-mail pings, conversations, and annoying eating noises that could disrupt your precious writing time. A little time management of the office hours can help here if you have the luxury of being able to modify when you can be in the office – for example, early in the morning or later in the evening (or even weekends) when fewer people are around. If your office is not conducive to writing because of intense traffic, you might seek the extreme option of requesting new accommodation, although the probability of being granted sanctuary on this basis alone will be low.

Another option, and one that I regularly choose when time permits, is to work from home. In fact, I have one colleague who writes almost entirely while sitting on a bed in his guest room (clearly when guests are not visiting), propped up with abundant pillows, curtains

---

[5] See also Chapter 18 for tips on how to manage stress-inducing e-mails.

drawn, and lights dimmed. He calls it his 'writing cave', and believe me, this strategy works extremely well for him (he is one of the most prolific and cited scientists in my field). While a dedicated at-home office might be preferable, it is not always possible in every living arrangement. I am in fact sitting on my own bed right now as I write these very words. The home option only works when you can minimise the distractions particular to homes; for example, trying to write during weekends or holidays when the children are home from school will probably not work as well. There are other many tantalising distractions in your own home, such as attending to chores that might appear attractive if you happen to be stuck in an overly tenacious bout of procrastination. Neither should you be anywhere near a television or telephone.

Others find that their favourite café offers the best writing venue; others might prefer a park bench, or the public library. Neither does your special writing place need to be a fixed address. Many of my colleagues do some of their best writing on the train or bus during the morning or evening commute, especially as a means to avoid eye contact with the millions of other commuters doing the same thing. If you happen to be travelling by aeroplane to a conference, for example, effective writing can be achieved by virtue of your enforced immobility. If you avoid the distractions of the in-flight entertainment system, then by all means take out that laptop computer and start writing.

MUSIC

I cannot possibly know how easy it will be for each of you to find that special writing place, because quiet places are becoming fewer and farther between in this increasingly bustling, interconnected world. One tool that can effectively mask distractions, especially noisy ones, is music. I consider my earphones to be an essential tool of the trade, for they allow me to 'tune out' as I 'tune in' to my favourite mood music. A little caution is required here. If the music is set too loud to mask the ambient noises that you are presently finding annoying, you might discover that your capacity to concentrate is reduced. The style of music is also important. When I am writing actual text, anything that could induce the slightest foot tapping or head banging tends to send me off into space; I prefer something light and instrumental in these circumstances, like Vivaldi, Mozart, or Miles Davis. On the contrary, if I am merely transcribing data, coding, analysing, or creating display items, then I tend to go more for heavy metal or electronica to set an intense pace. You will inevitably find some combination of music styles that works best for you.

DEADLINES

A great way to finish those languishing papers if all else fails is to impose fixed deadlines. I am not referring to deadlines that you might set yourself as some sort of goal for personal amelioration. Instead, I am referring to deadlines imposed by other people for unalterable events. One of the most immutable scientific events that imposes strict deadlines – for which the penalty of breach can cost real money or irretrievable damage to your reputation – is the scientific conference. If you deliberately set a deadline for writing a paper that is required before you can give your presentation on a specified date, I assure you that you will find the time to write it, even if it is the week before and during the entire journey to the conference venue. Nothing stimulates writing more than sheer panic.

SELF-IMPRISONMENT

Extreme situations sometimes call for extreme measures. It is something that I have only recently discovered, and I am slightly embarrassed to admit that it has taken me this long to incorporate into my repertoire of writing tools. In reality, this solution is not all that extreme, but I am surprised how infrequently most scientists seem to employ it. When you run out of time and place options for writing, often the only thing left to do is to send yourself into exile.

Find an unoccupied room at your institution (you might be required to book it in advance to avoid being disturbed), place your laptop on a table, draw the blinds, and shut and lock the doors. If you have to go to this extreme getaway, I also suggest that you do more than the minimal one- or two-hour writing block – an entire eight-hour day is better. While this can easily be done as a form of solitary confinement, I have found instead that including one other co-author with whom I work well is in fact more efficient. As a two-person team you can both encourage each other to write, much like a personal trainer at the gym encourages you to bench press just one more time. With a small team, it is also easier to assign particular tasks, such as one person writes the Methods while the other attends to the Introduction, and so on. In some senses, this exile acts like a mini-workshop (see the next section), so if you can keep the pace up for a few days, you will find that your paper writing advances more quickly than you might suspect it could.

VIVE LE WORKSHOP

Workshops can be some of the most efficient structures not only for writing, but also for doing big science that you alone could not achieve. First, I shall define what I think a 'workshop' is; for me, a workshop is a small group of like-minded scientists – all of whom possess different skills and specialities – who are brought together to achieve one goal. That goal is writing the superlative manuscript for publication, and this usually also means that discussion and analyses are also done at the same time. I am not referring to the bog-standard talk fests into which many so-called workshops descend. Workshops are not mini-conferences infected with motherhood statements and diatribes; neither are they soap boxes. It is my view that nothing can waste a scientist's precious time more than an ill-planned and aimless workshop. But with a little planning and some essential ingredients, you can turn a moderately good idea into something that can potentially shake the foundations of an entire scientific discipline.

So, what are these secret ingredients? Before you even decide that a workshop is necessary, first identify at least one big hypothesis that you suppose you can test with real data, even if those data might be difficult to obtain. I believe most scientists can think of at least a few burning hypotheses that they would love to test if the data were available. To acquire the necessary data, you might have to hire an assistant, or get a student to spend some time to put together the database (whether the data are collected *de novo* or collated from existing databases or individual papers). Sometimes this process can take months or even years. Getting the bulk of the necessary data collected well before the workshop actually takes place is essential. You do not want to waste your time trying to find the data during the workshop itself.

Now you should invite no more than ten to fifteen collaborators who can specialise in at least one component of the hypothesis to be tested. These can be people intimately familiar with the type of data, mathematicians or statisticians who have the latest analytical methods at their fingertips, taxonomists, and so forth. If you can see several dimensions to the big question, invite a specialist of each. However, it is not always required or even advisable just to choose the biggest names in the field as the specialists. A room full of crusty old professors will get about as much done as a room full of neophyte PhD students (i.e., not much). You will need a broad cross-section of experience and free time to make the workshop achieve its goals.

Just as good writing places and times are essential, a great workshop place and duration are even more so. My advice is to go to a galaxy far, far away. If you have your workshop next to your office, you will not be nearly as productive. Ideally, all participants should go to a neutral place far away from their work, family, and other day-to-day responsibilities. I can highly recommend field stations, wineries, retreats, and other beautiful places. You will inevitably 'waste' a bit of time refining your research questions when you begin, so one day is not long enough to get anything meaningful done. However, two weeks is probably too long because the participants will lose steam and focus by then. My advice is to go away for at least three days, but five to seven days is ideal.

Appointing a leader to keep the group to task is a good idea. Generally, this will be the workshop organiser, but it can be anyone in the group with the necessary leadership skills and vision. The leader will act as a guide throughout the process by assigning tasks and making sure people stay on top of them. If you are lucky enough to have administrative assistance to organise the sundries, then the leader can focus solely on the academic components; if not, then the leader usually takes the responsibility of hiring the venue, organising the catering, and booking the participants' travel.

As a group, you will collectively plan your output, much like I suggest you do when writing a paper (Chapter 4). Often workshops end up growing into unruly, multi-tentacled beasts if people are not kept to task. One great output is better than five mediocre ones. Here, the leader makes sure that everyone has a job to do and that this is clear. There is no better way to waste someone's time if they feel like a fifth wheel. Projectors and whiteboards are essential tools to set these tasks and evolve ideas. I also often organise break-out groups, which are like mini-workshops inside of the main one where specific tasks requiring the input of specific experts can be achieved most efficiently by working in groups of three to five.

You should strive to keep the participants as comfortable as possible. If they do not have to think about food, drink, or a nice place to sleep, they will focus more on the science at hand. I usually organise catering to come to the workshop, at least during the course of the day because it minimises disruption to momentum. Evening dinners are nicer in a private room in a restaurant where people can unwind and discuss things in comfort and peace. I also highly recommend providing good coffee and teas throughout the day. Also keep your group well-hydrated, although I generally recommend leaving the wine and beer until after beer o'clock (this time varies according to country), but make sure you provide enough tasty tipples in the evenings to relax the body and mind.

As I have already mentioned, internet access can be a huge distraction, so ideally you might consider banning everyone from accessing e-mails. In practice, this is nearly impossible to enforce, and it might inadvertently get up someone's nose. In fact, internet access is essential for finding those missing papers, augmenting datasets, and rapidly sharing files during the course of the event. Instead of placing a strict ban, I suggest that you politely encourage people to close down their browsers and e-mail programs while they work – they will have time to attend to those in the evenings.

I have never yet completed a workshop with a finalised manuscript in hand ready to submit to a journal, so usually some clear post-workshop deadlines must be set. Few things are more disappointing than gaining some good momentum during the workshop, then letting it languish in the months and years that follow. The leader should assign specific post-workshop tasks and deadlines to break-out group leaders. Be almost military about ensuring these are met on time.

While I have not always followed my own advice for running or participating in workshops, the best ones with the best outcomes invariably incorporate most of these aspects in their design. Of course, the perfect workshop takes time and money, so be strategic about how you invest your grant money (see also Chapter 12). It might seem expensive at the beginning (especially when you add up the travel, accommodation, alcohol, and catering), but it can be the most cost-effective way to achieve that published output required to acquit a research grant.

4   **Writing a Scientific Paper**

How difficult can it be to write a scientific paper? After all, a short introduction, a brief synopsis of the data collected and methods applied, a summary of the main results, and a discussion of their implications must be straightforward to cobble together once the hard work of experimentation, data collection, and analysis are complete, right? Were it that simple. I have never written a paper from start to finish like that. When I started writing them, it usually involved a few scribbles about some vaguely worded hypotheses, some rough, pixelated graphs pasted unceremoniously into the *Results*, and a badly formatted, partial reference list of other papers I should probably read before I submit the damn thing. I would then haphazardly add bits to each of the sections as I slowly fleshed out the Frankensteinian beast, finishing with a description of the *Methods*, and a bit of formatting. If my co-authors[1] were lucky, I would circulate semi-complete versions of the various sections to at least some of them over the course of the mishmash. As I suspect you will agree, this is probably not the most efficient way to write a scientific paper.

After many years of this variable, semi-random, and wholly unsatisfactory way of writing my papers, a long-term colleague of mine[2] with whom I shared lab directorship suggested that we should perhaps come up with a better system, especially for our neophyte doctoral students who were struggling to write their first papers. This is a common problem, and most supervisors probably get their collective paper-writing wisdom across to their students in dribs and drabs over the course of their degree. I imagine every supervisor has a

---

[1] See Chapter 5.
[2] Professor Barry Brook, now at the University of Tasmania in Hobart.

different style, emphasis, short-cut (or two), and focus when writing a paper, and students invariably pick up at least some of these passively as they progress.

The fact that this knowledge is not innate, nor is it in any way taught systematically in most undergraduate programmes, means that most supervisors must bleed heavily with the virtual red pen[3] on those first drafts. Bleeding is painful, for both the supervisor (who has to spend a lot of time and effort during the surgery) and the student (who has to suffer the indignity as well as clean up the mess). There have to be better ways, and indeed there are. Many fine books have been written about how to write scientific papers (e.g., 17–19), and I do encourage you eventually to read at least some of these, or their equivalents. But how many starting doctoral students sit down and read such books cover to cover? I can barely get them to read the basic statistics texts, let alone entire volumes on writing protocols.

Our method is therefore intended as a quick-and-dirty protocol that will at least get you started on the path to writing papers efficiently. It has proven to be remarkably simple, effective, and intuitive for those to whom I have recommended it. Although the protocol was designed for biological papers, it should apply to most scientific disciplines with some modest tweaking.

THE 12-STEP PROGRAMME

Step 1   Mind Map

Even before starting a new document file, you should set up a physical meeting with your co-authors to plan your paper together. Meetings should ideally be in person, but Skyping or any other type of video-conferencing can work. At this stage, not all eventual co-authors need attend, but you should have the major players in the (virtual) room. My preferred method is to arrange some chairs around a whiteboard (electronic or standard) so that you can quickly jot down ideas in front of everyone. As first author, you will make notes and collate

---

[3] Most likely using the track-changes function in programs such as Microsoft® Word.

discussion ideas, and because you are the one primarily responsible for deciding what goes into the paper, you should be holding the pen. Do not worry about self-censoring during this 'mind mapping' step.

What is jotted down is up to you, but I recommend starting with a few hypotheses, how to test them, what data or experiments are needed, as well as the 'main message' of the paper (see next step). You can also sketch out some 'display items' (figures and tables) that might be useful, as well as acknowledge any anticipated hurdles. This step should last no more than an hour or so.

Step 2   Main Message

This is one of the most important steps in the entire protocol, because it sets the limits within which you will restrict your data, analysis, and discussion. Write down the main message that you want to get across to the readership in twenty-five words or less (adhere to this limit; 26 words are too many). Although you might have multiple lines of evidence in your paper, you should still have only one main message. If you cannot think of just one, you are either not focussing enough, or you have more than one paper to write. It can help to think in terms of a 'pitch' – how would you 'sell' the idea to a potential funding agency in the time it takes to walk from your office to the tea room?

Step 3   Working Abstract

It might sound odd to write an *Abstract* before you have all your results and ideas nailed down, but trust me, this is great way to keep focus. A working *Abstract* should explicitly answer the following questions:

Why are you doing this? [context and aim]
What did you do? [methods]
What did you find? [core results – say something useful, i.e., no
    motherhood statements or deference to the main text]
What does this mean? [interpretation in context]

What is it good for? [application]: No one will bother to download and read your full paper (or cite it) if they do not find the *Abstract* interesting.

Step 4   Working Titles

Based on your main message and working *Abstract*, write down your best title. In fact, I like to write down two or three variants of the possible title just to hear if any of them have a better ring than the others. Titles are extremely important no matter how objective you might claim to be, for we are all susceptible to clever marketing. Titles are your first hook to get both editors and readers interested, and let me assure you, competition for readership is fierce. A good title should lure the casual browser to read further. In most cases, especially for primary data papers, give your main result in your title – this will be a direct link to your main message. Few people will bother to read your *Abstract* if your title is boring or lacks relevance.

Step 5   Distribute

It is now time to send your main message, working *Abstract*, and proposed title(s) to your co-authors, usually via e-mail. I usually set a deadline here of a maximum response time of just a few days because it should not take them long to read and critique. After their feedback, revise the three components and send them straight back to all co-authors without an explicit request to respond unless some debate arose after the first distribution. Iterate until at least the major debates cease, or until some form of compromise is reached. This is also the ideal time to settle any co-authorship contribution and order conflicts (see Chapter 5).

Step 6   Plan Display Items

This is the step where you decide exactly how many tables, figures, schematics, and drawings that you will ultimately include in the 'main text' of the paper, and what additional display items can be relegated to the supplementary information as appendices. Most journals these days allow for so-called 'supplementary' data, results, methods, and relevant text in a word-limitless set of appendices usually housed online by the scientific publishers (see also Chapter 11). In fact, the higher up the journal-impact ladder you go, the shorter the main text, the fewer the number of display items allowed, and the larger the length and complexity of the supplementary information. Many papers in journals like *Nature*, *Science*, and *Cell* have supplementary appendices that vastly exceed the length of the actual (main text) article, so decisions about what to include, and where, are essential at this stage.

To be able to do this, of course you will now have to have some idea about where you will be submitting your final manuscript, and most journal guidelines now state the maximum number of display items that they are willing to accept in the main article. If you really do not yet know to which journal you will likely submit, then I recommend a strict upper limit of six display items (any mix of figures, tables, etc.) to get you going. If you have more than six items, rank them in order of importance and move the lowest ranked ones to the supplementary information. Of course, you may have fewer than six.

Step 7   Create Display Items

Go ahead now and create the figures and tables to the best of your ability. Of course, this assumes that all of your analyses and results are complete, even though you might discover that there are some more things to analyse or more data to present down the track. Make sure you write out their captions in full now, because these will be used as reference points for writing the *Results* and *Discussion*. Ensure that each caption is stand-alone from the main text (i.e., it makes sense as it is and its interpretation does not depend on additional explanatory text).

Another trick to consider using when you have a lot of information to get across is to construct 'multi-panel' plots like the following example from one of my own recent papers (20).

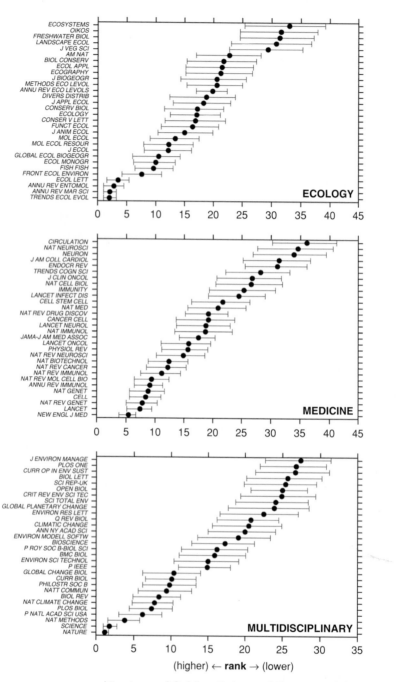

FIGURE 4.1 (Caption modified from Reference 20): Mean rank (± 95 per cent uncertainty limits) of the top 30 journals for two disparate biological disciplines: *Ecology and Medicine*, plus one *Multidisciplinary* (*caption cont.*)

Although I did not use them in this example, you can also add in a letter like 'a', 'b', 'c' in the upper left- or right-hand corner of each panel and refer to them specifically in the caption. Doing this over many figures can really save you a lot of space and maximise the amount of information presented in each figure.

Step 8   Circulate

It is now time again to send your choice of main-text display items to your co-authors, possibly embedded in the document that contains the title(s), main message, and *Abstract*. Once you receive their input (again, give them only a few days to respond) revise accordingly. Iterate until everyone is more or less happy with your selection and their presentation.

Step 9   Plan the Skeleton

This might seem like you might be overdoing things, but trust me, this step is very much worth the day or so it might take you to do it. Planning the paper's skeleton, or framework, requires careful thinking, but if done correctly it will appear as though the rest of the paper writes itself.

The first order of business is to decide on the length of the main text. Of course, this will vary from target journal to target journal (e.g., the maximum number of words and format for *Nature* is decidedly different to those for *PLoS One*). Unless you are targeting from the outset one of the high-impact 'magazine' journals like *Science*

Figure 4.1. (*caption cont.*)
theme. Journals are ordered by mean rank of five metrics: *Impact Factor* (Web of Knowledge), *Immediacy Index* (Web of Knowledge), *Source Normalised Impact per Paper* (Scopus), *SCImago Journal Rank* (Scopus), and Google's 5-year Hirsch's (h) index for journals ($h5$) $\div \log_{10}$(number of citable items).

Statistics were estimated using κ-resampling with 10,000 iterations, from a total sample of 100 journals for *Ecology and Medicine* and 50 journals for *Multidisciplinary* (see Reference 20 for details). Journal abbreviations follow the Web of Knowledge standard.

or *Nature*, it might not yet be clear to you which target journal will be the most appropriate (see Chapter 6). If you find yourself in this position, I recommend planning on approximately 20 double-spaced manuscript pages, or six printed pages (as it would appear in the journal). How long is this? Approximately 850 words of text per journal page, or 50 references, or four display items. So, roughly 3.5 pages of text, 1.5 pages of display items, and one page of references for an average[4] primary data paper. That is $850 \times 3.5 = 3000$ words of main text. If you need to write more (e.g., detailed methods), the additional words can go into the supplementary information that almost every scientific journal now allows.

It is important to stick as closely as you can to your set word limits, and it is a good idea to try to limit yourself to 50 main references, even if the journal to which you will be submitting allows more. It is likely that your co-authors, and certainly your reviewers, will suggest additional references to cite somewhere along the path to publication. It is decidedly easier to add references than to decide which ones to cut, so try to cite only the most important articles in the beginning.

Now estimate the relative size of each section (the standard *Introduction*, *Methods*, *Results*, *Discussion*, and if required – *Conclusion* sections). A rule of thumb is to split your 3000-word target into about 600, 900, 500, 800, and 200 words (or 20, 30, 17, 27, and 7 per cent), but it varies depending on how much you need to set up the context, and how many lines of evidence you are using, etc. Even with these variable scenarios in mind, it is surprising how often it works out to be these approximate proportions. If you are more relaxed about word length (say, for a review paper), then still aim to use these rough proportions.

For each section, you now need to plan the paragraphs, which should each be between 50 and 250 words in length. At this stage, however, do only the following: write out each paragraph's main

---

[4] You might have to modify this according to the average of your specific discipline.

message in fifteen words or less (similar to the concept of the paper's main message – remembering that each paragraph should be about only one thing). Then play around with the arrangement of the paragraphs until you are satisfied with the logical flow. If you wish, you can add to each paragraph some additional notes, key words, indications of references to cite, display items to refer to, etc. This helps to clarify or elaborate the 15-word main message.

Now your skeleton is nearly complete pending approval by your co-authors. You therefore need to circulate the entire skeleton to them and invite critical feedback. Emphasise to them that this is the appropriate time to address any major problems with logical flow, ideas, content, and thrust towards achieving the main message. As before, iterate the skeleton's revisions until all your co-authors are satisfied or at least until some reasonable consensus is reached. Often you will fail to receive any substantive feedback except 'This all looks good to me'. If this happens, do not sweat too much – it merely indicates that you have planned well up to now and that all co-authors feel included, like they have contributed, and that they are now too committed to raise any unpredicted fuss down the track.

Step 10 Write the Paragraphs

This is, somewhat counterintuitively, the easiest part of the paper-writing process. You can do this in any order you like because you know your structure and flow are already established. This is a great advantage, because some parts of a paper are inevitably easier to write than others (and getting more and more final text down is a psychological boost towards completion). This also punches through writer's block, and also permits you to work on discrete units of your paper to avoid mental burn-out. Some days you might feel like writing only one paragraph, whereas on others you might feel like you are in *the zone* and can write most of them in one long block of time. In general though, I recommend that you do not spend all day writing without a break. Instead, finish a distinct block and take a break before attempting another – you can do another task, meet a student, respond to some e-mails, take a walk, or do some more analysis or coding. It is also often a good idea to set goals for each day, such as completing three paragraphs and allotting one hour for each.

For me, the process of writing paragraphs is closely associated with the core and relevant literature that I have either not yet read, or have read too long ago to recall specific details accurately. I therefore tend to read or review the relevant literature as I write, especially for paragraphs in the *Introduction* and *Discussion* (one tends not to cite too many scientific articles in the *Results*, and there are usually only few methodological articles cited in the *Methods*). I add the references as I write, usually using some dedicated software[5] such as Endnote®, Mendeley®, RefWorks®, or the open-access Zotero (zotero.org), that embeds citations within my document on the fly.

There are of course other acceptable models, such as writing each paragraph completely first, adding reference 'placeholders' that you eventually update with the appropriate citations afterwards, or combining the cite-as-you-write model with placeholders when the appropriate references do not immediately spring to mind. Another exception to this general approach is when you write a review article that does not often follow the *Introduction, Methods, Results, Discussion* format. For reviews, referencing becomes part of the approach and analysis itself, so the citation model I described above does not often work as well (although the general article-writing approach does). While it is well beyond the scope of this book, there are some excellent references for how to write reviews (21), and *systematic reviews*[6] in particular (22).

---

[5] I tend to use Endnote® embedded within Microsoft Word®, merely because my institution has a licence for both and I have used them together for many years. I do not necessarily advocate using this specific combination – you should find something that works for you. However, I think it is important that you be able to embed the citations directly within your preferred document software, and that your co-authors mostly use the same system.

[6] "A systematic review, in contrast to a traditional narrative literature review, requires a clearly formulated research question, an extensive literature search that ideally includes relevant unpublished research findings, transparent study inclusion

Step 11   Revise the Abstract

Your paper is now almost ready to send to your co-authors, but you need to revise that old *Abstract* beforehand. In nearly all cases the process of writing the rest of the paper will have led to a change in logical order, hypotheses, or even results, and these changes need to be reflected in the new *Abstract*. But because you already have one written, it will be easier to revise rather than write from scratch.

Step 12   Circulate Again

Circulate this first complete draft of the manuscript to your co-authors and give them sufficient time for feedback. A good rule of thumb is something like two weeks as a deadline – substantially less than that and it might seem unreasonably restrictive; more than that might result in you losing momentum towards submission. You will most likely find that they will be happy to meet this deadline because they have already been embedded in the development of the manuscript from the outset, even if it is just to say 'great work' at each step when they were asked for feedback.

GENERAL ADVICE

In the *Results* and *Discussion* in particular, it is important to lead with your main points, followed by the next most-important, and so on. There is nothing wrong with a little repetition in your main message in the *Introduction* and *Discussion* – in fact, I encourage it. Strategic repetition reinforces your readers' understanding and is more likely to convince them that your article is clever and well-written. Of course, there is a balance to be struck here; too much repetition will annoy readers. You can avoid annoyance to some degree by using different words to say the same thing, as long as you are not clearly trying to hide the fact that you are repeating yourself by

and exclusion criteria, a quantitative synthesis of the data (normally by a **meta-analysis**), and interpretation of the results." Vetter *et al.* (2013). *Ecosphere* 4:1–24. doi:10.1890/ES13–00062.1

over-employing the built-in thesaurus. The first paragraph of the *Discussion* is ideally where you should place that strategic repetition of your main message.

While you are waiting for feedback from your co-authors at any step along the way, you can occupy yourself with some work on later steps, or you can certainly work on other manuscripts to maximise your writing efficiency. But never demand feedback from your co-authors; request it politely instead, and make the point that it is optional for any stage of the manuscript's construction *except* for the comments on the final draft. If they give none at a particular stage, it might just be that they are entirely happy with what you have produced.

It takes some discipline to follow these steps, but I encourage you to persist until it becomes second nature. Even if you start to think to yourself that the process is too repetitive or unnecessary, do not give up. This structured method does work well compared to a more random and disorganised approach that will ultimately waste more time and probably result in poorer-quality first drafts.

5   Sticky Subject of Authorship

If you have not yet written your first scientific paper, you might be scratching your head right now about the necessity of this chapter. Surely an author is an author – someone who *writes* the paper. Were it that simple. Few scientific papers these days are 'written' by a single author, because so little science is done without the contribution of many others. You might have certain people collecting data in the field, others who organise the laboratory experiments, some who work primarily to identify specimens, specialists to run expensive analytical equipment, database managers, and expert statisticians to analyse the data. Then there are usually still others who planned the entire research project and obtained the grant money that allowed it all to go ahead in the first place. There are also typically collaborators from other laboratories or organisations, the person in the tea room with whom you might have discussed and embellished the original idea, and the poor student who tirelessly collected all the relevant papers in the area and summarised the current state of knowledge on the subject for their final-year essay.

Often there are herds of potentially deserving individuals who at least loosely fit the definition of 'author' and could conceivably be listed as such on the eventual papers. While it is certainly an exception to the normal number of authors (and this number varies considerably amongst disciplines), I myself have co-authored a paper with 255 others (23), and the current record for the largest number of authors on a single paper is 5154 (24). The problem is that there are no hard rules about when to include someone as a co-author, and when just to thank them in the dedicated *Acknowledgements* section. It can be a confusing and potentially dangerous game to play – I have even had to appear in front of a committee consisting of the Deputy Vice

49

Chancellors for Research of two universities to decide the order of authors on a paper I had co-written with four other researchers.[1] Such co-authorship wars have strained relationships between collaborators, led to sackings, ended friendships, and even destroyed marriages.

Despite having therefore developed a few guidelines in my lab regarding the authorship of articles, I have come to realise that each article requires its own finessing each time authorship is in question. Authorship issues can be divided into two main categories: (i) who to include as a co-author, and once the list of co-authors has been decided, (ii) the order in which they should be listed. Before discussing the issues related to Category i, it is prudent to declare that there are probably as many conventions as there are scientific disciplines, and each discipline's most general conventions differ across the scientific spectrum. I am sure if you asked ten people about what they considered appropriate, you could conceivably hear ten different answers. That said, I do still think there are some general good-behaviour guidelines on authorship that one should strive to follow, all of which are based on my own or my colleagues' experiences (both good and awful).

So, who to include? It seems like a simple question superficially because clearly if someone contributed to writing a peer-reviewed article, he/she should be listed as a co-author. The problem really does not concern the main author (the person who did most of the actual composition), because it is clear here who that will be in almost every case (my university 'judicial' hearing notwithstanding). In most circumstances, this also happens to be the lead author, but more on that below. The question should really apply then to those individuals whose effort was more modest in the production of the final paper.

Strictly speaking, an 'author' should write words; but how many words do they need to write before being included? Would ten suffice, or at least 10 per cent? You can see why this is in itself a sticky subject because there are no established or accepted thresholds. As

---

[1] In that case my argument was deemed valid and I was given the position of first author, because, as I demonstrated to the committee, I had done all of the analysis and written over 95 per cent of the words.

I mentioned above, science generally requires much more than just writing words: for most papers, there are experiments to design, grants to obtain to fund them, data to collect, analysis and modelling to be done, figures and tables to prepare, and finally, words to write. I admit that I have co-authored many papers where I have done mainly one of those things (analysis, data collection, etc.), but I can also hold my hand over my heart and state that I have contributed more than a good deal to the actual writing of the paper in all circumstances where I have been listed as a co-author (the amount of which depends entirely on the lead author's writing capacity – see Chapter 2).

It is my opinion that if someone contributed to any of the components required to complete the article, then they should be given the opportunity to claim co-authorship. The only proviso to that policy relates to two of these sub-components: (a) collecting data, and (b) funding the research. In most cases, it is pretty clear that collecting data takes a lot of time and effort, so if someone who collected the data you need passes it along to you, you should probably offer that person authorship for any articles arising. That said, there are many freely available databases out there, so it would be completely unfeasible to offer authorship to all the players involved. In a legal sense, if the data are publicly funded and available online, you have every right to publish papers based on them without necessarily including the custodians as co-authors (although you are obliged to cite and acknowledge them). Most databases give clear directions in this matter, so do read these. If it concerns a colleague with a less formalised database, then authorship is a little more straightforward. My rule of thumb? If in doubt, offer the data custodian a place on the author team.

Scientists spend a huge amount of time begging (errhm, ... requesting ...) funding to do their research, which *inter alia* requires original thought, collaboration, novelty, good communication skills, impressive track records, and a certain flair for impressing funding agencies (see Chapter 12). If someone funded the work that led to your paper, you should probably also consider asking them to co-author your work. However, if it is the 500th paper arising from a major grant with scores of collaborators, it might not always be sensible or necessary to do so. The main guideline here is that you should always communicate openly and sincerely with anyone who might be affected by this issue. Another problem that can rear its ugly head is when someone is paid to put together a dataset or to collect data. In most cases I know, the person specifically employed to do so generally *is not* expected to become a co-author, unless she/he demonstrates exceptional effort, novelty, independence, and foresight. I have employed more than a few research assistants who meet all these latter criteria and more, and so I automatically include them as co-authors on any papers arising from 'their' datasets.

All these considerations, complexities, social contracts, and expectations in mind, the best advice I can give on authorship is this: it rarely hurts you to include someone of marginal contribution, but it can really hurt you to exclude someone who at least thinks she/he deserves authorship. Engrave that sentiment into your frontal lobe for posterity, because following it will make a huge difference to your long-term publication success.

Now for the stickiest issue of all – author order. Why should the order of authors matter at all, for surely if their names all appear in the list then it should not matter who is first and who is last? But it does matter, because there is an unspoken assumption that some authors contributed differently to others, and that the order somehow reflects this (but not in the intuitive sense of the first contributing the most, the second, the second-most, ..., and the last, the least). Relative author contribution is – explicitly or implicitly – an important component in ranking researchers for job interviews, grant application assessments, promotions, and awards. I have even heard of situations where some assessment panels ignore every author but the first (or the first and last – see more below) when tallying a scientist's publication performance.

The main reason author order is so sticky is that there are even fewer rules of thumb to follow, even after deciding on the final number of authors. I will describe what I tend to do, but it really depends on the discipline and the culture of the lab to which you belong. Of course, first authors should be the ones that tend to do the most work, and especially the actual writing. There are times when this is not always the case, but it is so standard that there should not be many situations where it is difficult to determine who this is.

In my discipline anyway, there is an increasingly recognised trend that the second-most important position in the author order is the last one. This tradition probably goes back to medical research laboratories where the lab head generally occupied the final position in the list of authors (whether or not she/he actually wrote any of the paper). The culture was strengthened when the British Research Assessment Exercise ratified the practice by placing emphasis almost exclusively on first- and last-author positions for papers counting towards one's 'official' track record. In Australia, the practice is becoming more and more mainstream, hence one of the reasons that I tend to take the last position in most papers coming out of my lab and for which I arguably have the second-most important contribution to overall authorship.

Talking to colleagues in France, USA, and elsewhere, I am not sure the practice is as strong, but you should at least be aware that it exists. After the 'last position' rule, the next most important position is certainly the second author. After that, it tends to be a case of diminishing contributions up to the penultimate author. In cases where this is not necessarily clear, and certainly when there are many (i.e., >10) authors, the ones between position 2 and n-1 are often listed alphabetically by surname. The latter trick is a good way to avoid infighting amongst co-authors. Sometimes even the entire author list is alphabetised (oh, the lucky ones with surnames beginning with 'A').

Increasingly, there are situations where two or more authors have equal contributions. It is perfectly acceptable, and now even conventional, that such authors are indicated (usually by some sort of superscript symbol) as having contributed equally. This often includes the first two authors, or the last two, but I have even seen it applied to positions two and three, or to even more than two authors. For anything but the lead and second author, however, the paper will nearly always be referred to as SURNAME1 et al.,[2] so take the 'equal contribution' convention with a grain of salt.

In conclusion, make sure you discuss with all your authors any issues that could arise *before* they do. Heading off any future problems by discussing them well beforehand is a good way to avoid woes later on. If you follow my paper-writing protocol from Chapter 4 and circulate to all potential authors from the outset of the idea, you will also avoid many of the worst problems before they can occur.

[2] et al. is an abbreviation of the Latin *et alia*, which means 'and others' (i.e., the rest of the co-authors). Although this is probably not a fair way to represent an author's contribution, it is done largely to avoid long, clumsy lists of names in the main text of the paper.

6   Where and What to Publish

The progress of science is built on the foundations of previous research – we take the flame of our predecessors and pass it faithfully to the next generation of scientists, and that 'flame' principally takes the form of the written word. So, it has always been; so it shall (ideally) always be. Hence, the backbone of science is an ever-growing brick wall of published evidence to which subsequent research can refer, and at times challenge. Scientific articles have more recently also started playing another role: as metrics of the progress of research projects and of the 'quality' of researchers and institutions (25). Regardless of the pros and cons of this secondary function, boosted by a parallel increase in the number of researchers (26), there has been an enormous increase in the number of peer-reviewed scientific articles. There are now well over 50 million peer-reviewed scientific articles in existence (27), with an increase of 8–9 per cent each year over the past several decades (28). This means that over one and a half million new articles are published each year across all scientific disciplines (27).

With these millions of papers, let alone all the books, book chapters, and reports now published annually, the challenge therefore for any scientist is to make herself known and heard (read) in this extremely competitive market. So far, I have mentioned the peer-reviewed article as the main venue for the dissemination of your scientific results, and I will always maintain that this should be your main publishing outlet. The question is how much I mean 'main', and what other formats you should consider as your career progresses. Of course, these other formats to which I am primarily referring are authored and edited books, book chapters, and reports. I am not, however, going to discuss the so-called 'non-academic' writing outlets such as blogs, opinion editorials, and magazine articles here (for that, and related mass-media outlets for your writing, see Chapters 21 and 22). There is no ideal proportion of effort one should dedicate to each of these categories, and even within categories it can be dreadfully difficult to choose just where to send your scientific manuscripts.

THE 'RIGHT' JOURNAL

Let us start with the main output category for any serious scientist – the peer-reviewed journal article. In many ways, you will live or die as a scientist depending on how well you do in this particular arena – the so-called 'publish or perish' paradigm that, for all its injustices, inefficiencies, and incongruences, is more than ever an overarching driver of scientific endeavour (25). However, merely publishing in any old journal will not necessarily get you as far as you could in terms of the metrics of 'effectiveness' outlined in Chapter 1. Indeed, the standard on which any quantitative measure of a scientist's performance is increasingly based is the number of citations you acquire for your published articles (29). As the real-estate cliché goes, it's all about 'location, location, location' – the scientific version is simply 'citation, citation, citation'.

Unless you are not a scientist at all, or still a fledgling considering a pathway towards becoming one, you will undoubtedly have at least heard someone mention the concept of 'journal metrics', and all of them are based to some degree on citations. Human beings love to rank themselves and others, the things they make, and the institutions to which they belong, so it is a natural expectation that scientific journals are ranked as well. But as you progress your way through your career, you will also notice that scientists, and to perhaps a greater extent, their administrators and overseers, are apparently obsessed by these ranking metrics. But why are journal metrics and the rankings they imply so in-demand? Despite many people loathing the entire concept of citation-based journal metrics, we scientists, our administrators, granting agencies, award committees, and promotion panellists use them with such merciless frequency that our academic fates are intimately bound to the 'quality' of the journals in which we publish.

What is 'citation-based'? *Citations* are merely the number of times a particular body of scientific work (be that a peer-reviewed journal article, a book chapter, a book, a report, etc.) is mentioned officially in another body of scientific work. For example, if in a paper I am writing I *cite* another paper to justify why I posed a particular hypothesis for my current research, then when my paper is accepted for publication, the article I cited receives one citation. If I cite one of my own previously published papers, this is called an 'auto-' or 'self-' citation.[1] Over the course of a paper's lifetime (which is for all intents and purposes, infinite, but in practice tends to be a few decades at most), a published paper can receive many thousands of citations, although most receive substantially fewer than that (and some are never cited at all). Clearly, a paper that makes a bit of a scientific splash tends to receive more citations than another that is a little more run-of-the-mill, so if the editors of a particular peer-reviewed journal choose to publish papers that, on average, obtain more citations than those published in another journal, then a citation-based ranking would rank the former journal higher.

I am certainly not the first to suggest that journal *quality* cannot be fully captured by some formulation of the number of citations its papers receive; 'quality' is an elusive characteristic that includes *inter alia* things like speed of publication, fairness of the review process, prevalence of gate-keeping,[2] reputation of the editors (i.e., globally

---

[1] While it is inevitable that you will cite your previously published work because it tends to save time and avoids repeating information, some authors abuse the practice by citing themselves to the exclusion of more appropriate papers written by others. It is for this reason that many metrics exclude auto-citations in their calculations.

[2] *Gate-keeping* is the nasty tendency for those in positions of power, be they long-recognised experts in their fields acting as reviewers or established journal editors, to make it difficult for new people to break into their particular field of science. Perhaps the most prevalent form of gate-keeping is powerful men attempting to exclude women scientists from positions of power, although perhaps that is starting to change for the better (see Chapter 15). Other types include established scientists limiting the rise of early-career researchers, or highly discipline-specific researchers

recognised expertise), writing style, discipline reputation, longevity, cost, specialisation, open-access options, and even its 'look'. But it would be impossible and highly subjective to include all of these aspects into a single 'quality' metric, although one could conceivably rank journals according to one or several of those features. 'Reputation' is perhaps the most quantitative characteristic when measured as citations, so we academics have chosen the lowest-hanging fruit and built our quality-ranking universe around them (30).

The oldest and still most widely used citation-based journal metric is the Impact Factor produced by a for-profit company (Thomson Reuters) called the Institute of Scientific Information (ISI®) *Web of Science* (webofknowledge.com). It might seem strange that a publicly listed corporation generates a subscription-based[3] metric to rank journals, but when you realise just how profiteering the academic publishing industry can be (Chapter 8), this will not surprise you in the slightest (although it should still make you angry). The ISI Impact Factor is calculated as the average number of times articles from the journal published in the past two years have been cited in the *Journal Citation Reports* year, the latter meaning the last year for which ISI tabulated the number of citations. Many have criticised the Impact Factor because of its many flaws (31–33), including that it does not compare well among disciplines (34, 35), it tends to increase over time regardless of actual journal performance (35, 36), and the methods behind its calculation are not transparent (in particular, what types of articles are counted for tabulating total citations). Nonetheless, the Impact Factor is now entrenched in the psyche of researchers and has arguably changed the dynamics of journal assessment and bibliometrics more than any other single method [15].

Despite its established dominance, the Impact Factor has, however, many competitors that are all to some degree based on citation

---

precluding related research from other disciplines modifying their limited world view.

[3] This means that either you or your institution has to pay a rather substantial subscription fee to access these and other services provided by ISI.

data. Today, these include some of ISI's other metrics such as the five-year average Impact Factor (i.e., the average Impact Factor over the last five years), the Immediacy Index (i.e., the average number of times an article is cited in the year it is published), Cited Half-life (i.e., the median age of articles cited by the journal in the *Journal Citation Reports* year), Eigenfactor Score (which is based on the number of times articles from the journal published in the past five years have been cited in the *Journal Citation Reports* year, but it also considers which journals have contributed these citations, such that highly cited journals will influence the network more than lesser-cited journals; thus, the Eigenfactor Score is not influenced by journal self-citation[4]), and Article Influence Score (calculated by dividing a journal's Eigenfactor Score by the number of articles in the journal, normalised as a fraction of all articles in all publications) (37).

The academic publishing giant Elsevier also produces some metrics based on the Scopus® citation database[5] (another paid-subscription service), including the Source Normalized Impact per Paper (i.e., the ratio of a journal's citation count per paper and the citation potential in its subject field; this is based on the average length of reference lists in a field to determine the field-specific probability of being cited, which is used to correct for differences between subject fields) (38), Impact per Publication[6] (i.e., the ratio of citations in a year to papers published in the three previous years, divided by the number of papers published in those same years) (38), and SCImago Journal Rank (i.e., a measure of scientific influence of scholarly journals that accounts for both the number of citations received by a journal and the importance or prestige of the journals from where such citations

---

[4] In this case, *self-citation* is the practice of citing papers from the very same journal in which the article is published – some journals have been caught and duly punished for encouraging their authors to cite papers from their journal more than others.

[5] journalmetrics.com.

[6] Take note that the Impact per Publication score is no longer calculated by Scopus; instead, the Centre for Science and Technology Studies provides it at journalindicators.com.

come. It is a variant of the eigenvector centrality measure used in network theory (39, 40). Most recently (December 2016), Elsevier released a new ranking metric also based on Scopus data, called *CiteScore* that is intended to challenge the dominance of the Impact Factor. The metric includes three years of citation data and all document types citing the journal's articles, and it is apparently (so they claim) more transparent than other metrics. The web giant Google has also produced citation-based journal rankings (scholar.google.com), including the 5-year Hirsch-type[7] (41) index (h) (42) and its median (43). The Google h-index is the largest number h such that h articles published over the last five years have at least h citations each.

You would be forgiven for assuming that because all of these different metrics are based on some variant of actual citation data, they should give roughly the same journal rankings (even though the actual scores might be on incomparable scales). Unfortunately, this is not the case, for different metrics can deliver vastly different rankings for individual journals (43, 44). As such, no single metric can be viewed as ideal because some tend to overestimate citations (e.g., Google Scholar) (45), while others underestimate them (e.g., Web of Science- and Scopus-based metrics) (33, 46). How therefore could any single individual deem which journals are higher-ranked, and hence, which have higher citation potentials?

Before I proceed with a possible solution to this dilemma, I have to backtrack a little and revisit the main theme of this section: where should one send one's manuscripts? First, it behoves any serious scientist that all other things being equal, maximising the citation potential of your work should be a priority. In other words, you should want lots of other scientists to cite your work for the reasons I gave earlier (i.e., more citations lead to more grants, more awards, more promotions, etc.). So logically, you should aim to submit your manuscripts

---

[7] Jorge Hirsch, a physicist at the University of California, San Diego, originally devised what is now known as the 'h-index' for individual researchers. Like its more recent journal version, an individual's h-index is the number of papers with citation number higher or equal to h.

to the journals that promise (at least, statistically) to give your articles more citations.

Taken to its logical conclusion, this strategy would ultimately see you submit every single one of your manuscripts to the top-ranked (mainly generalist) journals like *Nature*, *Science*, *Cell*, etc. Unless you are a natural genius and innate science superstar (most of us are not), this strategy would unfortunately backfire and result in far *fewer* citations overall. This is because of the extremely high rejection rate of these journals and the fact that few of your articles are unlikely to be Earth-shattering enough to warrant publication in these venues. This is not to suggest, even for a moment, that your work is not valuable and important; it simply means that you are unlikely to publish many articles in these top-notch journals over the course of your career.

So, a cleverer submission strategy is required. Assuming again that we want to maximise each article's citation potential, a middle-of-the-road approach is often more successful. This means that you should save your most wow!-factor articles for the big journals, but that you should submit your more routine manuscripts to journals occupying the more median ranks. If you wanted simply to guarantee an easy ride during the reviewing process (see Chapter 7 for more on running the gauntlet of peer review), you would instead submit all of your manuscripts to the lowest-ranking journals. But this would almost certainly lead to few citations overall, so that is clearly not the best strategy either. My recommendation therefore is that for each manuscript, identify the journals to which you think it has some (reasonable) chance of being accepted, also taking into consideration each journal's theme, reputation, cost, open-access options, editors, and punctuality, and submit to the journal that occupies the top-ranked position. If you are rejected and cannot appeal the decision (see Chapter 7), then move to the next journal down the list, and so on. Sometimes you will get it right and your manuscript will sail past the gauntlet of the highest-ranked journal in your list, but many more times it will fail. This might mean a few more rejections overall, but it will also maximise your citation potential in the long run.

Now back to journal ranks, and the problem I outlined above regarding the inconsistency in different ranking metrics. How can you reasonably determine what is 'top-ranked' out of any list of target or thematically grouped journals? Given the inconsistencies and the inherent biases that all journal-ranking metrics embody, I eventually decided to take matters into my own hands and brave the murky world of bibliometrics (statistical analysis of publication patterns). With my long-time colleague, Professor Barry Brook now at the University of Tasmania, we devised an entirely new way of reconciling the different ranking systems that exist. This system is a more comprehensive, although still admittedly simple, approach to estimate the relative ranks of journals from any selection one would care to cobble together. We also included a rank-placement resampler to estimate the uncertainty associated with each rank (20).

Without going into too much detail (you can read all about it in the paper (20) itself[8]), our approach took five of the most well-known, least inter-correlated, and easily accessible citation-based metrics (Impact Factor, Immediacy Index, Google Scholar 5-year h-index, Source Normalized Impact per Paper, and SCImago Journal Rank), ranked them, then calculated the mean rank per journal. My use of the terms *relative* and *within* the sample are important here – journal metrics should only ever be considered as indices of relative, average-citation performance from within a discipline-specific or personally selected sample of journals (such as the 'submission candidates' suggested above). By itself, the value of any particular journal citation metric is largely meaningless. The procedure therefore gives you a relative ranked list from one (highest-ranked) to n (lowest ranked of n candidate journals) that you can then use for your submission-sequence strategy. It avoids the arbitrary scales and biases of any individual metric while simultaneously providing a measure of rank uncertainty. As I indicated previously, a citation-based ranking,

---

[8] See what I am doing here? I am citing myself in the hope that someone else will also cite our paper. Nudge, nudge; wink, wink.

even one that coalesces different metrics, is still only one aspect you should use to choose where to submit; ultimately, your and your co-authors' expert opinion will also come into play.

I have a few final comments regarding the drive to maximise your articles' citation potential. First of all, you should be aware that even if you manage to publish in the highest-ranking journals, it is by no means a guarantee that your paper will be cited a lot. While these journals tend to have a high *average* citation rate per paper they publish, the median is rather smaller. This means that many of their published papers ultimately receive few – if any – citations at all, whereas only a handful receive many hundreds or even thousands within the first few years of publication (47, 48). The converse is also true of lower-ranked journals. An article published in such journals is not necessarily doomed to a citation netherworld, because it might appeal widely to a selection of specialists, or hit some other unanticipated chord among your colleagues. I must also reiterate that – if only for my own conscience – one should try to avoid the temptation of making citations your all-consuming focus. While I hope you agree by now that citations are important, if they remain your only preoccupation, you will potentially compromise the quality of your scientific work. I make this warning because the temptation to play the citation game to the exclusion of all other considerations might lead to poor choices, such as cutting corners on methodological detail, overly sensationalist language (49) that glosses over uncertainty, or the ill-advised truncation of words necessary to describe complex scientific concepts. Imagine how superficial the body of scientific knowledge would be if all results were published only in magazine-style *Science* or *Nature* formats; I shudder to think what we would be missing.

NOT JUST JOURNAL ARTICLES

Without question, most of your scientific output should be in the form of peer-reviewed journal articles, but there are many other ways to publish your work. The principal reason journal articles should be your mainstay is that at least traditionally, citations were counted exclusively from the journal category. However, today citations from books, book chapters, theses, and reports are increasingly counted (e.g., by Google Scholar and Elsevier's *CiteScore*), so the foundations for that excuse are now starting to crumble. Regardless, it is likely that at least for the foreseeable future, most scientists will be gauged primarily by their track record of peer-reviewed journal articles, although a smaller proportion of their work can still be published elsewhere.

Let us start with the most common alternative venue – the book chapter. I have always been wary of these, despite having written quite a few myself, as well as edited books where I have commissioned others to write chapters. While the reasons for such trepidation likely vary among disciplines, my experience suggests that book chapters can be something of a curse. Even though now generally counted, book chapters are on average less likely to be cited than primary journal articles. This is because most books cost money to purchase, so unless you or your institution's library happens to buy them, they are less likely to be read by your colleagues. Another strike against book chapter citations is that many scientists have the impression – justified or not – that chapters are generally of a slightly weaker quality than journal articles; in other words, the results reported in chapters are deemed to be less reliable. While this is by no means always the case, scientists writing book chapters have a tendency to rush their work through, or report results that are less solid, than they would for journal articles, because scientists do not want to 'waste' their best work on a mere chapter tucked away in an expensive and difficult-to-access book. Indeed, some compiled scientific books are no better than a loosely affiliated collection of second-rate works from scientists who believe that their results would be difficult to publish elsewhere.

My appraisal might sound a bit jaded, but it is only an overall impression. Clearly, there are times when book chapters are ideal; for example, specialised topics that amass results from experts can become pivotal tomes in their respective fields. Some methodologically focussed books in particular often end up being indispensable laboratory resources, and so they can command a wide readership and citations. Budding scientists are often asked directly to contribute book chapters either because they have yet to publish much of their work in journals, or because their mentors recommend that they represent good venues to practice scientific writing while avoiding the worst of the nastiness embedded within the peer-review system (Chapter 7). Whatever the reasons for or against writing book chapters, my advice remains that they should represent only a small proportion of your scientific publications.

Once upon a time the doctoral or master thesis was considered a perfectly acceptable final format for scientific publication by early-career research students. This is categorically no longer the case. In almost all scientific disciplines today, the thesis is increasingly considered to be a mere formality for obtaining a postgraduate degree, and that the contents therein should already be published in, under consideration for, or about to be submitted to, peer-reviewed journals. There are still vestiges of the old-school approach to the doctoral thesis in particular, that it represents a complete and highly interconnected tome of chapters not unlike a book. However, more academics today are instructing their students instead to compile collections of stand-alone (albeit, related) papers that are submission-ready for journals upon completion. I would even go so far as to predict the eventual demise of the traditional thesis in favour of the latter. It is for this reason mainly that few people today cite theses, and so they should almost never be the final publication goal.

Reports are the worst of the lot for alternative venues of scientific output, principally because there is no guarantee of quality control. By 'quality control', I mean of course peer review through a recognised process (Chapters 7 and 8). Even book chapters can go through dubious peer review at the whim of the particular editors, but reports are generally far less reliable. While some scientific institutions such as non-governmental organisations, think tanks, and international organisations embrace the report as an important publishing medium, the scientific community as a whole generally favours a primary (peer-reviewed journal) source when citing a supporting body of work. That is not to say that reports cannot play a role; reports are often commissioned by paying customers that support a scientist's research, but if that is where the results only ever end up, the chances are that they are unlikely to receive many citations. I recommend much caution here.

The scientific book – distinguished from the book chapter as a book written by one or a few authors in its entirety – is the last major, non-journal category that I will discuss in this section. While I wager that a fair few of you reading this now have many science books on your shelves, the prospect of writing one probably appears more than just a little daunting. Take it from me, it is not a trivial endeavour to write even a short book like this one. Science books can take on many different roles: they can (*i*) be popularised versions of a large, complex body of primary scientific results; (*ii*) elaborate technical or methodological detail that just cannot be packed neatly into even a long-format journal article; (*iii*) expand conceptually on a complex topic that would be otherwise difficult or impossible to do within journal articles; or (*iv*) represent an entire career's worth of information and insight that the next generation of scientists might find useful.[9] So, writing a book does have its place, but it is not normally the sort of task one embraces early in one's career. Many non-science scholars write books almost exclusively (e.g., researchers in most of the humanities), but that is certainly not the model adopted by most scientists. My recommendation is that you should only consider writing a book once you have a respectable track record of journal articles, and that you feel comfortable that there is a need, a sizeable audience, and an appetite for the prospective book's contents.

One final word about publishing preprints. Preprints are almost exclusively manuscripts of articles that have not yet been through

---

[9] I hope that the current book will fall into this last category.

the peer-review process. It might seem odd to some of you that a scientist would 'publish' (i.e., post) their unverified results on a publicly accessible server before they could benefit from having it published outright in a reputable, peer-reviewed journal, or it might be unfathomable to others that you would not. Physicists, mathematicians, and computer scientists have traditionally adopted the latter approach by electing to publish many or most of their manuscripts first on the site known as arXiv (arXiv.org[10]). The main idea here is that a pre-reviewed manuscript can attract comments and informal reviews from peers to improve the presentation, reliability, and readability of the article prior to formal submission to a peer-reviewed journal. In theory, it is a great idea because it should provide many opportunities to improve scientific rigour, although other disciplines such as biology have been warier of the practice and today still only sparingly use this, and a more discipline-specific, pre-print service (bioRxiv.org). Whether you choose to use the service depends on your personal philosophy, the fashion within your discipline, and your personal or colleagues' experience; personally, I am ambivalent about the whole affair.

WHAT TO PUBLISH

Where to publish your results is one question, but what *type* of article to publish is another matter entirely. There are no rules here, and I suspect each scientific discipline, and even each scientist, will have their own recommended guidelines. That acknowledged, there are some general principles to keep in mind when writing up your work. Considering our main objective of writing journal articles, there are many different types to consider: standard data articles,[11] models, reviews, methods, comments, opinion editorials, conceptual

---

[10] An alternative to arXiv.org is viXra.org (yes, the latter is indeed a palindrome of the former) that publishes preprints from all scientific disciplines.

[11] Approximately following the IMRD format (*Introduction, Methods, Results, Discussion*).

treatises, and horizon scans.[12] While a good review can collect many citations, if a scientist only ever wrote these, she might not be viewed favourably as a 'proper' scientist by her peers. The same applies to comments, opinion editorials, and conceptual pieces – these are based on existing knowledge and so do not necessarily represent science in its truest definition (testing hypotheses with data). Standard data articles, model descriptions, and methods papers should probably therefore make up the bulk of your published track record, with the first category dominating. In essence, you should strive to create a strong foundation of data papers that test hypotheses, followed by a good deal of methods and reviews (or modelling papers, should that be your particular forte). Comments and opinion editorials can be wielded effectively to your advantage, but they should represent only a small proportion of your total work.

[12] Becoming more common, these articles allow scientists to speculate about what might be just around the corner in terms of new technologies, ideas, possibilities, and discoveries.

7 The Publishing Battle

Unless your path to becoming a scientist has only just begun and you have not yet actually formulated and tested any hypotheses, then you might be of the mistaken assumption that science is a nice, linear, logical process of informed questioning and elegant experimentation leading to unambiguous conclusions. If you are already a seasoned scientist, you will appreciate the sarcastic humour of the previous sentence. While every scientist certainly aims to emulate that particular ideal, it almost never happens as cleanly as initially intended. Failure, at least of the temporary kind, is therefore ironically a state you will necessarily have to get used to if you are to succeed as a scientist. But such failures (let us call them 'temporary setbacks' from now on), do not just apply to the scientific process itself, because laboratory or field setbacks are really part of the journey towards scientific discovery.

I will instead tackle a subject in this chapter that is rarely discussed formally, but it will certainly challenge your mental health during the course of your scientific career. I am referring of course to the often uncomfortable and typically adversarial process of peer review. On the surface, the peer-review system sounds harmless enough – indeed, it is the very essence of science today – but I can assure you that few other aspects of a scientist's day-to-day activities will elicit as much indignation, wrath, hatred, stress, worry, despair, and feelings of gross inadequacy as this essential process. The main reason these feelings will haunt you at some point during the submission of your manuscripts is because of one word: REJECTED.

REJECTION

I cannot even recall the number of times I have witnessed a student's face figuratively melt as he shuffles into my office to show me a journal's rejection letter, which is now preceded almost always by an e-mail forwarded to my inbox. Unfortunately, we scientists can be real bastards to each other, and no other interaction brings out that tendency more than peer review. Few people outside the sciences have even the remotest inkling about how nasty it can get. Use any cliché or metaphor you want, it applies: dog-eat-dog, survival of the fittest, jugular-slicing ninjas, or brain-eating zombies in lab coats. While students and other early-career scientists tend to take these hits the hardest, I want to impart a little wisdom from my own experience as well as from my well-established and successful colleagues. Instead of seeing the inevitable rejections as a mark of failure, they should instead be viewed as an asset. I shall explain.

Of course, no one, no matter how experienced, likes to have a manuscript rejected. People generally hate to be on the receiving end of any criticism, and scientists are certainly no different in that way. Many reviews can be harsh and unfair; many reviewers 'miss the point' or are just plain nasty. Ask any PhD student after receiving the referees' comments on her first paper – most often it involves an outright rejection, typically accompanied by some caring and supportive words like 'fail', 'flawed', and 'nonsense'. It does not improve either as you progress through your career – you just become numb to the pain and soldier on. It could be argued that some *Schadenfreude*-afflicted reviewers even find some sort of twisted pleasure by stomping on others; regardless, I argue that if you are *not* getting rejected, you are probably not trying hard enough.

In keeping with my recommendation in Chapter 6 to aim high when selecting candidate journals to which you are contemplating submission, it is inevitable that you will be rejected outright many times after the first attempt. Sometimes you can counter this negative decision via an appeal (see below), but more often than not the rejection is final no matter what you could argue or modify. So, your only recourse is to move on to a lower-ranked journal (see Chapter 6). If you consistently submit to low-ranked journals, you would obviously receive far fewer rejections during the course of your scientific career, but you would also probably minimise the number of citations arising from your work as a consequence. That is not a good idea.

So, your manuscript has been REJECTED. What now? The first thing to remember is that *you and your colleagues* have not been rejected, only your *manuscript* has. This might seem obvious as you read these words, but nearly everyone – save the chronically narcissistic – goes through some feelings of self-doubt and inadequacy following a rejection letter. At this point it is essential to remind yourself that your capacity to do science is not being judged here; rather, the most likely explanation is that given your strategy to maximise your paper's citation potential, you have probably just overshot the target journal. What this really means is that the editor and/or reviewers have judged that your paper is not likely to gain as many citations as they think papers in their journal should. Look closely at the rejection letter – does it say anything about '...lacking novelty...'?

The 'lacking novelty' excuse is an old trick that you will see in most rejections not explicitly highlighting any technical flaws. Truly 'novel' science is in fact rather rare, because so many scientists have come before you and asked similar questions, posed similar hypotheses, collected similar datasets, and applied similar approaches. Not only is 'novelty' rare, it is also somewhat irrelevant because for any scientific conclusion to persist, it must be repeated using different data and analyses from different labs and by other people. Hypotheses that are not rejected must have support, and if enough of them lead to the same general conclusion, then theories can be developed, from which eventually laws evolve if contradictory findings do not reset the entire process. So, what 'lacking novelty' really means is 'we do not think your paper will get as many citations as we would like'.

There is not much you can do about that judgement unless you resubmit an essentially different manuscript, so normally if the rejection is based primarily on the 'novelty' excuse and not on some misunderstanding or clearly biased assessment, your only recourse is to submit the manuscript somewhere else. Unless you deem the process to be rigged from the start (I would have my doubts), you should also think about making at least the major changes suggested by either the editor or reviewers who spent some time looking over your work. Generally speaking, specific science disciplines are microcosms, so if your paper continually gets rejected from one journal to the next, chances are it is being seen by some of the same people. If all you do is repackage the manuscript for each journal without changing any of the text, you might be inadvertently limiting your chances with each submission by appearing to ignore everyone's advice. If you do find that journals are continually shutting doors in your face, take a close look at your manuscript. Perhaps it needs a rebuild and a reset before it will see the light of day.

The take-home message then is that rejection is not an indication of a scientist's worth or capacity, it is merely an indication that you are shooting high and want to succeed, provided you follow through and eventually get the work published somewhere.

MAJOR REVISION

Counter-intuitively, the most challenging outcome for many scientists is not outright rejection, but 'major revision'. Even as recent as a decade ago, 'major revision' meant exactly that – the editor (and reviewers) wanted you to make substantial changes to the manuscript – be they compositional, analytical, numerical or otherwise – before it could be considered for publication in the journal. These days however, 'major revision' has been recoded as 'reject with an invitation to resubmit'. The simple reason the words (but not the meaning or implications) have changed is that nearly all journals attempt to inflate their attractiveness by promising quick turnaround times (i.e., the average time it takes a paper to be published after its initial submission). By 'rejecting with an invitation to resubmit' (translation: *major revision*), journals play another dirty trick and reset the turn-around clock so that their handling statistics appear to be much faster than they really are. Do not believe the hype – it is just false advertising.

Irrespective of what they now call it, *major revision* is clearly a foot in the door, although it is still a long way off a guaranteed publication. In an ideal world, you proceed to make the required changes and resubmit the manuscript for (usually) another round of assessments by typically the same reviewers. But the world is not ideal, so the first mental hurdle is dealing with the venom spat upon the virtual pages of your manuscript. Unless it concerns a list of friendly and constructive criticism, I urge you to avoid reading too much of the assessments the very moment you receive them. My reasoning is that when you eventually get down into the details and absorb the nastiest of comments, your first feelings will be shock and probably a fair degree of hurt. If you are like me, those feelings are soon replaced by anger and disbelief, followed by some despair. If I am sounding overly dramatic, I apologise, but I cannot avoid travelling an emotional rollercoaster each time this happens. Of course, given this has happened to me quite a few times by now, I have learned just to roll with it, but the feelings are always there nonetheless. My strategy is therefore to skim the comments just to taste the general flavour of the main criticisms, and then I close the e-mail message and think about something else for a few days. In so doing, I prepare myself slowly over those few days for the time that I will eventually have to respond to each nasty comment. I find that by resisting the temptation to leap straight into the ring that I am better prepared for the ensuing fight once I find the necessary time and mental state to broach the revision. The wait also softens my responses and reduces the risk of offending the editor or even the reviewer with inappropriate and impolite retorts.

The only danger of such a strategy is that putting off the revision because it will be painful can become a self-fulfilling endeavour; the longer you delay, the more difficult you will find it to start the revision. This negative feedback can easily turn a few days into a few weeks or a few months, by which time the window of opportunity for revision might disappear entirely. I am ashamed to admit that this has happened to me more than once, so I recommend that you wait no more than a few days, and certainly no more than a week or two, to start the revision. Another handy trick to lighten the load is to assign particular tasks to your specialist co-authors, such as attending to questions regarding methodology or analysis. Task assignments should also ideally be accompanied by deadlines so that your revision progress is not overly delayed by just one or two laggards.

RESPONDING TO REVIEWERS

Just like there are many styles to writing scientific manuscripts, there are also many ways to respond to a set of criticisms and suggestions from reviewers. Likewise, many people and organisations have compiled lists of what to do, and what not to do, in a response to reviews of your manuscript.[1] It clearly is a personal choice, but from my own experience as an author, reviewer, editor, and of what I have gleaned from the myriad suggestions available online, there are a few golden rules about how to respond:

(*i*) After you have calmed down a little (see previous section), it is essential that you remain polite throughout the process. Irrespective of how stupid, unfair, mean-spirited, or just plain lazy the reviewers might appear to you, do not stoop to their level and fire back with defensive, snarky comments. Neither must you ever blame the editor for even the worst types of reviews, because you will do yourself no favours at all by offending the main person who will decide your manuscript's fate.

(*ii*) If the decision requires substantive changes (see previous section), then it is often a good idea to summarise in a paragraph or two (or using point form) the major ways that you have listened to your reviewers and improved your manuscript accordingly. You can place this brief summary just before your point-by-point responses (see below), or in a separate 'letter' to the editor as per specific journal guidelines.

(*iii*) Make the editor's job as easy as possible. By this I mean that you should address each reviewer's critique or suggestion in order, whether that be

---

[1] To see what I mean, just type 'response to reviewer comments' or similar phrase into your favourite search engine and behold the plethora of available advice.

the order in which you received them, or grouped thematically if several reviewers highlight the same issues. I recommend copying the entire set of comments into a separate document, and then culling them to their bare-bones, basic message; there is no need to repeat every single word the reviewers wrote in their full assessments. After the cull, address each point immediately after it appears, differentiating the reviewer's comment and your response by font (e.g., italicised comments, normal-font response), colour (but this is not recommended for colour-blind scientists like me), or by a leading 'RESPONSE:...' or something similar in each response. As an editor, I want to be able to understand the essence of the reviewer's issue at a glance, and then concentrate on your concise, yet comprehensive response to it.

(iv) Unless the reviewers recommend only superficial or minor changes, do not automatically do everything they suggest. Reviewers are not omniscient, so they will often suggest inappropriate things, or request unreasonable, new analyses or experiments. If you can defend your approach, or can clearly identify why a particular suggestion is unwarranted, then by all means, do so. However, do not use this advice as an excuse to be lazy; if you cannot honestly demonstrate why the reviewer's suggestion is inferior or unsupported, then just follow the recommendation.

(v) For every remaining critique or suggestion, demonstrate to the editor that you have changed at least *something* in the manuscript, and identify where the changes now reside (either by new line numbers or some other navigational pointer). If you are required to rephrase important parts of the text, you can simply copy the revised sentence, paragraph, or (brief) section into the response letter to show the editor what a well-behaved and conscientious author you are. On the other hand, if a particular comment requires little change (e.g., adding a reference, re-wording a sentence, changing terminology, etc.), you can probably get away with something like 'we have now made this change', or even just 'done'.

(vi) There are no page limits for response letters, so feel free to add as much detail as is necessary without fatiguing the editor. It is a balancing act to be sure – insufficient detail or failing to respond to a particular concern will result in either another round of reviews or an outright rejection, whereas too much verbiage can bore the editor and distract her from the important business of accepting your manuscript. It is not that rare to have response letters that exceed the length of the manuscript itself, especially for the magazine-style, high-impact journals.

(*vii*) This advice can differ from that of others, but as an editor I am not typically overjoyed by reading copious 'thank you' statements or gratuitous repetition of the reviewer's compliments by the authors. Instead, stick to addressing the main critiques and do not ingratiate yourself.

In summary, the overall impression that the editors and reviewers[2] must have after reading your revision and response letter is that you have taken their advice seriously and made a substantial effort to accommodate their expert suggestions into the revised version. If they experience any other emotion than this, chances are greater that your manuscript will be rejected.

APPEALS

Up to here I have only mentioned one avenue following rejection – submitting your manuscript somewhere else. But there is a second option, and that is to appeal against the decision. This is a delicate subject that requires some reflection, so I have devoted an entire section to it. Early in my career, I believed the appeal process to be a waste of time. Having made one or two of them to no avail, and then having been on the receiving end of many appeals as a journal editor myself, I thought that it would be a rare occasion indeed when an appeal actually led to a reversal of the final decision. It turns out that I was very wrong, but not in terms of simple functional probability that you might be thinking. Ironically, the harder it is to get a paper published in a journal, the higher the likelihood that an appeal following rejection will lead to a favourable outcome for the submitting authors. What? Let me explain.

---

[2] In nearly all cases of 'major revision', at least one, and mostly likely all, of the original reviewers will see your responses to their comments. It is therefore essential that you avoid offending them.

Appeals are generally a port of last call with manuscript submission, but if you invoke that right each time you are rejected, you will quickly learn how futile an exercise it is, and you would most likely develop a reputation amongst your peers for being more than just a little annoying. Appeals should *only* be used when there is a clear pathway for reconciliation that typically rests on the shoulders of an awful or obviously biased review with inappropriate, personal overtones. The caveat about *novelty* notwithstanding, if you can convince the editor politely that there has been a major injustice or error on the part of one or more reviewers, then you have something to work with for an appeal. Indeed, when I receive a review that is little more than a badly disguised personal attack (alas, it does happen from time to time) and it becomes the fulcrum upon which a rejection rests, I usually rejoice because I can use it as a strong case for an appeal. Other cases where the reviewers are not necessarily out of line, but might be dreadfully confused or inexperienced (or just incorrect), can also give you a reason to appeal. Make sure that there is a glaring error on the part of the reviewers though before you even contemplate appealing.

I casually added a vital adverb above that might have escaped your attention: *politely*. Under no circumstances whatsoever should you deviate from the path of utter decorum, dignity, and politeness, no matter how maddeningly snide and uncouth the reviewer chooses to reveal displeasure with you or your manuscript. Bite your lip, avoid sarcasm, and be as sickly sweet as possible, while still exposing the flaws in the review and decision. Editors are (sub-)[3] human too, and so they will respond negatively to insulting language and obvious anger. Instead, take the higher moral ground and outline your arguments logically and without even the faintest scent of recrimination.

Now back to that seemingly flagrant contradiction that appeals are more successful in higher-ranked journals. If I had not been

[3] I could not resist that little jest because many authors, especially those on the receiving end of a rejection and who have never before been editors themselves, are wont to caricaturise editors as pitiless demons. Generally speaking, they (we) are not.

involved personally in more than just a few occasions where out-right rejections from the top-of-the-top journals were successfully appealed, I would not believe it myself. Top-ranked journals are in fact top-ranked because, as I have discussed in Chapter 6, they tend to publish manuscripts that have a higher-than-average (usually, much higher) probability of being cited. In a way, these editors of the elite journals could be said to *expect* appeals against rejection for that very reason; they want you to try to convince them that your paper will ultimately attract many citations (and not just because it is a pile of unmitigated excrement that deserves a comprehensive thrashing from your peers). If you have submitted to a high-ranked journal what is obviously, at least to you, some of your best work, then you should by all means consider an appeal if you have a good reason to do so. Do not be bashful and think that the editor will be annoyed and dismiss all of your future submissions out of hand just because you had the gall to appeal this one time. Stubbornness, with the appropriate foundations, has an important role to play in getting your papers published in the best journals.

ROCKING THE SCIENTIFIC BOAT

One thing that has simultaneously amused, disheartened, angered, and outraged me over the course of my academic career is how anyone in their right mind could even suggest that scientists band together into some sort of conspiracy to dupe the masses. While this tired accusation is most commonly made about climate scientists because of the extreme politicisation of that branch of physics, it applies across nearly every facet of the sciences whenever someone does not like what one of us says. First, it is essential to recognise that we are just not that organised. While I have yet to forget to wear my trousers to work (I am inclined to think that it will happen eventually), I am still far away from the superlative of 'organised'. Such is the life of the academic given the many roles we must assume. More importantly, the idea that a conspiracy could form among scientists ignores one of the most fundamental components of scientific progress – dissension. And my, can we dissent!

Yes, the scientific approach is one where successive lines of evidence testing hypotheses are eventually amassed into a concept, then perhaps a rule of thumb. If the rule of thumb stands against the scrutiny of countless studies, then it might eventually become a 'theory'. Some theories even make it to become the hallowed 'law', but that is rare indeed. The 'scrutiny' part of the process is where perhaps the second-nastiest behaviour of scientists comes to bear (the cruellest being, of course, peer review), for we regularly, and with a certain smugness, love to attack our peers' work in print[4]. In fact, much of the scientific literature essentially reduces to 'they are wrong because...'. I am certainly not complaining about all of this antagonism, for without it, *science* would cease to be *science*. We are bound inevitably to confrontation because it ultimately adds to the rigour of our scientific conclusions. It is merely the process that is sometimes painful, and of which every new scientist should be acutely aware.

One of the most adversarial types of scientific confrontation takes the form of the 'comment' and 'response'. What typically happens is that you and your colleagues write a paper and succeed in publishing it in a reputable, peer-reviewed journal. Most often nothing untoward happens after that, but eventually you might hear through a colleague or online somewhere that someone (or some group) deems your work to be problematic. Within usually three to twelve months after the publication of your paper, you might then receive an e-mail from the journal's editor that so-and-so has written a commentary critiquing your paper that they intend to publish in the very same journal. You are then most often invited to respond to the commentary within a certain restrictive format. After reading the dissenting views of your peers, and then carefully responding to the critique, both

---

[4] And now, in the courts. In one of the most ill-advised and spectacularly anti-science gestures I have witnessed yet, a researcher at a certain high-profile university in the USA has attempted to sue the authors critiquing his work for US$10 million. Google it.

'comment' and 'response' are usually published in the same issue of the journal. Your colleagues then have the opportunity to read the lunge and riposte with all of the fervour of children watching two of their fellow pupils beat nine colours of shit out of each other in the school yard.

Although the exchange should be congenial, typically it is anything but that. Just like a rejection during peer review, a 'comment' is often rather venomous even after being tempered by a diplomatic editor. Even the thickest-skinned and most battle-hardened scientists rarely relish receiving a critique, because it is just not fun to be on the receiving end of the attack – I have seen colleagues crumple in despair upon reading the first critique of their work. I have been involved in many of these mêlées over the years, and I suspect there are many more to come. The two things I have realised about all this is that (a) you cannot get away with even a small amount of unsupported speculation (i.e., bullshit) – someone will catch you out (and will go for your soul even if your work is rock-solid), and (b) if you are NOT receiving this kind of attention, you should be asking yourself why you are in science at all.

The boundaries of scientific knowledge are rarely pushed outward without some kind of fight. Sure, testing, re-testing, and re-testing some more are essential components of science, but if all we ever did was ratify or refine what we already understand well, scientific understanding would not progress quickly, if at all. If you really want to make a splash in science – and I am convinced most of you reading this do – you need to elicit stronger reactions from your peers, and not always in the form of mere admiration. In fact, I contend that there is no better indication that you are making your peers think twice about a particular scientific concept than receiving a 'comment' on your work. This is certainly no justification for unsupported sensationalism or deliberate deception, for those are not part of the scientific process. But challenging paradigms, or even the accepted *status quo*, are where the real meat of scientific understanding can be found.

So, when the inevitable attack of your paper arrives unceremoniously and without warning in your inbox, treat it like a negative review. As I have recommended above, you might consider putting it aside for a few days before absorbing the contents. Then after the first depressing emotions begin to subside, you and your colleagues can craft your riposte. But the difference here is that instead of dreading the possibility that others have taken issue with your work, you should rejoice that at least someone has noticed your paper and taken the time to read it, and that their disagreement is evidence of an impact. In fact, I contend that if you never receive comments on your papers, you might consider asking yourself if your work is too restrained or uninteresting. Critiques are, in my opinion, badges of honour in science that should be worn with pride.

8 Reviewing Scientific Papers

Apart from publishing your first peer-reviewed paper – whether it is in *Nature* or the *Journal of Disciplinary Parochialism*[1] – receiving your first request to review a manuscript is one of the best indications that you have finally been recognised as a 'real' scientist by your peers. Finally, someone important is acknowledging that you are an 'expert' in your field and that your opinions and critiques are important. You deserve to feel proud when this happens. Of course, reviewing is the backbone of the scientific process, because it is the main component of science's pursuit of subjectivity reduction. No other human endeavour can claim likewise.[2]

It is therefore essential to take the reviewing process seriously, even if we do so only from the entirely selfish perspective that if we did not, no one would seriously review our own work. Indeed, reviewing your peers' manuscripts is much more than an altruistic effort to advance human knowledge – it is at the very least a survival mechanism. If you did not participate in the process, or you only ever provided terrible[3] reviews, sooner or later if you would acquire a reputation that would most likely come back to haunt your attempts to publish your own work in the future. Despite this necessity, we are currently embedded within a highly unjust system that exploits the mostly free labour of reviewers (as well as authors and editors) to the advantage of a few, extremely wealthy publishing companies. Caught between the proverbial rock and hard place, we must comply despite our intellectual slavery (more on this later in the chapter).

---

[1] Not a real journal, obviously.

[2] See the final chapter.

[3] Not 'terrible' in the sense of negative, but just poor, incomplete, biased, or otherwise bad.

But I digress. Just like there are many different ways to write a scientific paper (Chapter 4), most scientists develop their own approaches for reviewing their colleagues' work. However, that does not indemnify the many awful reviewers in existence who are tasked to provide the quality control that is an essential element of the scientific process. Every single scientist alive today (and I wager rather a lot of dead ones, if only they could lament from beyond the grave) have at least a few stories to tell about the times when they received appallingly written, aggressive, mean-spirited, incomplete, incorrect, or just plainly idiotic reviews from so-called peers. The experience is even more difficult to swallow when the editor appears oblivious to the absurdity of such reviews and instead relies on them to make a decision about the quality of our submitted work. Such outcomes are unfortunately inevitable from time to time, but by following a few good reviewing guidelines, you can avoid perpetuating the atrociousness.

A review request generally comes from a 'subject' or 'handling' editor of a particular peer-reviewed journal (although occasionally it can come directly from the Editor-in-Chief) via an automated e-mail that tends to lack any sort of personalised text like the one below:[4]

Dear Professor Bradshaw:

I invite you to review 'Some really boring title' (manuscript number 12345, abstract below) that was recently submitted for publication in the *Journal of Disciplinary Parochialism*.

Please let me know as soon as possible if you will be able to review this manuscript. Clicking the appropriate link below will register your reply automatically with our online manuscript submission and review system. If you agree to review this manuscript, you will be notified via e-mail about how to access the manuscript and reviewer instructions in your *Reviewer Centre*.

---

[4] Names changed to protect the guilty.

Expert reviewers such as yourself are critical to the success of the *Journal of Disciplinary Parochialism*, which is committed to the rapid publication of the dreariest research in scientific parochialism. If you are unable to review this manuscript within 14 days, please e-mail me with recommendations for other expert reviewers who might be appropriate for this manuscript.

Thank you for considering this invitation. I look forward to hearing from you soon. The Editorial Team at *Journal of Disciplinary Parochialism* looks forward to your future involvement in assessing manuscripts for publication suitability.

Sincerely,
Professor Really BigName-Inmyfield
Subject Editor, *Journal of Disciplinary Parochialism*

Most such invitations include the *Abstract* as well as the list of the manuscript's authors. Such reviews are therefore known as 'single blind' because only one side of the party (the authors) will be known for certain by the other (the reviewers). There is an increasing tendency for reviews to be 'double blind' (neither the authors nor reviewers are indicated explicitly in the assessment material[5]), although at least in my field this is still rare. Personally, I see no disadvantage to double-blind reviews, but the debate regarding its relative merits and drawbacks is fierce (50, 51). As a rule too, I generally avoid identifying myself to the authors as a reviewer of their manuscript. It might appear to be noble and magnanimous to do so, but bruised egos can carry disproportionately large grudges for far too long. While some probably disagree with me, my advice therefore is to avoid the potential hassle and keep your reviews as anonymous as possible.

If the authors are identified, the first thing to consider is your relationship to them, even before you contemplate your availability,

---

[5] However, with a little experience, it is usually not much of a challenge to discover who the authors are, either by excessive self-citation, well-established writing styles, or adherence to particular schools of thought. It is also possible to guess the identity of the blinded reviewers, even if accuracy is not as high.

suitability, or interest. It is *essential* to avoid conflicts of interest at this stage, even if the idea of vengeance for past misdeeds is tantalising. In short, if you have an axe to grind with one or many of the listed authors in the invitation-to-review e-mail, my strongest advice is to decline politely. In the long run, you will not help yourself at all by trying to quash other scientists' work via an ill-advised reviewing rampage that futilely attempts to prevent publication. You will only anger the authors, who will in all likelihood be able to guess your identity eventually, and subsequently attack you mercilessly as a result the next time they get the chance to review your work. What comes around, goes around.

If you do accept to review the manuscript after contemplating your capacity to be impartial, read the entire paper once. This might seem an obvious and unnecessary point to raise, but I am convinced that many people who have 'reviewed' my own papers (and subsequently recommended rejection) have not gone past the *Abstract*. I cannot estimate how many times I have had to point out to an editor that the reviewer apparently missed an essential component of the manuscript that was (at least to me) rather clearly stated already. I will admit in these days of extended supplementary appendices that often exceed the length of the main paper (for the entirely unsatisfactory reason[6] that journal space is 'limited'), important information can get a little buried in the heap. There are also times where I have not stated the information as clearly as I should have, which means a casual read might miss the information in question. Those issues aside, the best way to approach a new manuscript is to read it through entirely at least once without focussing too much on bad writing, grammar issues, spelling mistakes, or other minor things, no matter how much they might annoy you.

After you have read the paper for the first time, put it aside and wait at least half a day before you attempt the full review, much

---

[6] It is demonstrably *not* limiting anymore – the decline in hard-copy journals and the ubiquity of online publication means that this argument holds little water. The internet is not full.

like I suggested when dealing with your peers' reviews of your own papers. This will (*i*) give you time to reflect on the main message and approach, and (*ii*) allow you to calm down if the authors have happened to claim something that angers or annoys you. This delay also gives you the time to contemplate the paper's potential importance to your field, and to prioritise how you might approach the actual review once you get started.

After you have made the decision to start the real review, I recommend that you ignore the *Abstract* entirely at the beginning. I treat the *Abstract* (and the title, for that matter) of most papers as advertisements rather than real scientific 'meat'. Even scientists tend to over-emphasise or even sensationalise their findings in the *Abstract*. I also submit that most people do not follow a logical pathway to compose their Abstracts either (see Chapter 4), so they are often the most poorly written components of submitted manuscripts. And as the great writer Mark Twain summarised beautifully, 'I did not have time to write a short letter, so I wrote a long one instead', meaning that concise writing can be difficult to achieve. By definition, the *Abstract* must be concise, which is a challenge to most writers at the best of times. It is therefore best to ignore this until right near the end of your review to increase the chances of providing a (more) objective appraisal.

As I recommended before, as you digest the manuscript text you might discover that the path of least resistance is to get hot and bothered about the little things, such as spelling mistakes, bad grammar, incorrect terminology, and poor construction. Despite arguing that these things are terribly important in your own writing (Chapter 2), it is not the reviewer's task to write (or rewrite) someone else's manuscript. Instead, a reviewer should apply most expertise and time to identify the 'big problems'. In the interest of efficiency, avoid wasting your and the authors' precious time, and go instead straight to the heart of the manuscript to determine whether it deserves a more in-depth appraisal. If there is clearly a flawed approach, an obvious and major misinterpretation or worse, such as fraudulent and evidence-free claims, then there is no need to bother dissecting the small problems. However, do not use this advice as an excuse to provide a one-line review just because you are lazy or nasty; rather, use it to avoid spending days to review a paper that has little hope of passing to the next stage anyway.

Despite this approach, it is still sometimes necessary to seek advice about a paper. While you are morally and legally obliged to ensure author anonymity to your peers, in my view it is entirely acceptable and justifiable to consult other specialists if you are not sure of a particular component of a manuscript. If the technique being presented is entirely foreign to you, or if you know someone who can tell you right away whether or not there is a problem, it is fine to ask for advice. However, you must not pass along the entire manuscript or reveal the authors' identity, but by all means you can discuss the general points with others if it will give you a more objective insight. If you do find yourself requiring rather a lot of advice (unless you have never before reviewed a manuscript), you might ask yourself if accepting to review was a good idea in the first place. You are also permitted, if not encouraged, to engage with the editor if you are in doubt about anything. Most editors are (mostly) human, and so they will often appreciate a discussion over e-mail, Skype, or telephone if necessary when a manuscript is contentious or borderline. As an editor myself, I appreciate the frank discussion with expert reviewers in circumstances when I am having a difficult time deciding on what to do with a problematic manuscript.

There is also a wise way to critique the problems, be they fatal or mere flesh wounds. As the mother's favourite proverb goes, if you cannot say something nice, do not say it all. Although it is not prudent to follow this advice to the letter during a review, my point is that you should strive to be constructive rather than destructive in your wording. It really helps no one at all if you are cruel. Even if you think, or say out loud for that matter, that the paper is 'rubbish', 'flawed', 'not worthy', or some such other unpleasantness, avoid the temptation to write out that sentiment. It is much more erudite and helpful to point out ways to improve the problems after you have identified them (politely). Nothing discourages authors more from improving their science (after all, is not that the point?) than receiving a negative criticism with no offer of a way forward. In other words, be mature and adult about it. If the person or persons whose work you are reviewing have been nasty to you before, try to take the moral higher ground.

Remember too that it is perfectly acceptable to disagree. Avoid the logical fallacy that agreement is equal to validation, for we are all human beings and subjectivity creeps into almost every paper at least to some degree, especially when discussing the uncertainty associated with the interpretation of results or the mechanisms underlying observed phenomena. It is alright if you disagree with the interpretation, provided the methods appear sound, the results are presented clearly and objectively, and the hypotheses are well-reasoned. A difference of opinion should not be used as the sole basis for recommended rejection. By all means, you should state your disagreement (again, politely), but conceding a difference of opinion as opposed to fact in such cases can be nobler and morally defensible. Make certain too that you understand your role – a reviewer is not an editor. Language such as 'I reject this manuscript...', or 'As written this is unsuitable for publication' is beyond your remit and your rights. A good reviewer should instead stick to the technical aspects of the appraisal and avoid making judgement calls about what the journal should or should not 'accept'.

In terms of organising your comments, I always recommend starting with the positives. I have not always followed this advice myself, but experience has taught me that a review that launches immediately into that is wrong with the paper is far more demoralising than one that attempts to point out at least some of its strengths. I am not recommending that you adopt a patronising tone by attempting to soften the sometimes-deadly blow (the assessment equivalent of pat on the head and condescending smile, if you will). However, I contend that all submitted manuscripts have at least one redeeming feature; even if there is only one, find it and acknowledge it straight away.

As you conscientiously accumulate all the good and bad elements of the manuscript you are reviewing, it can sometimes be tempting to write entire treatises about all the various elements of a complex study. But just like providing a single-line comment would be useless to both the editor (who makes the decision to accept, reject or more likely, request extensive modifications) and the author (who receives little or no guidance on how to improve), a tome of rants and overly prescriptive recommendations is just as unhelpful. As I previously stated, it is not the reviewer's job to rewrite the paper, even if she is the micromanaging, parental type. Instead, provide reasonable recommendations about how the authors can labour to complete revisions. In total then, I recommend spending no more than one or two days working on a review. Spending more than this on a review is probably just wasting your time, and most likely approaching that dangerously long treatise you should avoid.

As I have discussed in Chapter 2, I am something of a stickler for good grammar and terminology, hence my reviews typically end with a few pointers to the authors about how they can improve theirs. I make no attempt to hide my opinion that these are overlooked elements of the clarity of scientific writing, and that too few authors pay attention to the importance of clear language. That said, I always place these minor comments at the end of the review and I *never* use them as the sole or main basis for recommending rejection to the editor, except in the rare cases where the writing is so appalling and unintelligible that it completely masks the scientific message.

If I am guilty of any reviewing hypocrisy, it most certainly lies in breaking the next rule: being timely. While it might appear to the beginner reviewer that she can choose the pace at which she submits her reviews, it is decidedly bad form as well as self-defeating to hand in late reviews. Everyone else unequivocally hates delayed reviews – the authors, the editors, and the other reviewers – and most of them will know who you are. I agree that there is not always the greatest incentive to finish a review on time (I elaborate on this in the next section of this chapter), but I appeal to your sense of impatience with unacceptable handling times of your own manuscripts that late reviews are to be avoided.

Finally, I implore you not to do too many reviews. While I am certain that all scientists would enjoy reducing the handling times and editors would appreciate an overabundant pool of potential reviewers to choose, I have noticed a tendency amongst early-career researchers to spend too much time reviewing and not enough time writing their own papers. There is no hard-and-fast rule, but I recommend reviewing no more than one to two papers per month to avoid excessive time lost (others have different opinions) (52). It is a delicate balance, and it will not always be a consistent frequency (some months you might need to review five manuscripts; other months, none at all), but doing too much of one or the other will hurt you in the long run.

KNOWLEDGE SLAVERY

I hinted previously that there are some serious problems with the way science is quality controlled and published. For all my recommendations about quality, fairness, and timeliness to which every reviewer should aspire, I wish I could say that the scientific publishing industry did the same. At risk of the editor of this book cancelling my contract, I will admit though that there are some components of the science publishing world that are a little more ethical. Books are a good example, for I am paid a certain royalty[7] for every book sold, and generally speaking, a publishing company will also pay a modest sum to experts who read the book proposal and to others who review the final product. On the other hand, I contend that publishing articles in nearly all peer-reviewed journals amounts to a form of intellectual slavery. I

---

[7] I could argue that most royalties are not high enough, but I am not what I would label a 'greedy' person. If you are greedy for monetary recompense, I contend that science is probably not your ideal career choice.

brazenly defend my use of the word 'slavery' here, for how else would you describe a business where the product (scientific results) is produced by others (scientists) for free, is assessed for quality by others (reviewers) for free, is commissioned, overviewed and selected by yet others (editors) for free, and then sold back to the very same scientists and the rest of the world's knowledge consumers at exorbitant prices? To make matters worse, most scientists have absolutely no idea how much their institutions pay for these subscriptions, so there is little consumer scrutiny passed from researcher to administrator. In 2015, Jason Schmitt of Clarkson University in Potsdam, New York quoted Brian Nosek, Director of the Center for Open Science (cos.io), to sum up the situation (53):

> Academic publishing is the perfect business model to make a lot of money. You have the producer and consumer as the same person: the researcher. And the researcher has no idea how much anything costs. I, as the researcher, produce the scholarship and I want it to have the biggest impact possible and so what I care about is the prestige of the journal and how many people read it. Once it is finally accepted, since it is so hard to get acceptances, I am so delighted that I will sign anything – send me a form and I will sign it. I have no idea I have signed over my copyright or what implications that has – nor do I care, because it has no impact on me. The reward is the publication.

Some journals go even beyond this sort of profiteering and also inflict 'page charges' of hundreds to thousands of US dollars on the authors for the privilege of having their work appear in that journal.

I am not just grumpy about what many might assume to be a specialised and irrelevant sector of the economy, because it is in fact an industry worth many billions of dollars annually. In fact, one of the biggest corporations, Reed-Elsevier, made over £1.8 billion (nearly US$2.8 billion) in adjusted operating profit in 2015 (54). Other major publishing companies like Wiley-Blackwell, Springer, Taylor & Francis, and Sage Publications, which with Reed-Elsevier collectively published more than half of all the academic papers published in 2013 (55), make many billions in profit each year as well: Wiley-Blackwell took in US$965 million in revenue in 2016 (56), Springer had a 2012 revenue of US$1.26 billion (57), and Sage Publications had a 2015 profit of $585 million (58). If these sounds like huge sums, they are, for the reasons I outlined in the paragraph above. Worldwide, scientific publishing has huge profit margins; for example, in 2015 Reed-Elsevier's profit margin was a monstrous 37 per cent, with Springer following closely at 35 per cent and Sage Publications at 28.2 per cent (58, 59). For comparison, Apple – one of the most profitable companies in the world[8] – had a profit margin of 29 per cent, Google's was 25 per cent, mining giant Rio Tinto's was 23 per cent, and the car maker BMW's profit was 10 per cent (59).

Many other scientists have of course noticed this profiteering, and in protest have stopped submitting manuscripts to the journals of particular companies, or have refused to review or edit for them. Back in 2012, there was a huge push to boycott Reed-Elsevier in particular because of its exorbitant journal subscription prices (60), and while certain scientists have maintained their disgust by refusing to do any work for that company, the main movement has largely lost steam. The latest iteration of mass protest is a petition[9] by Finnish academics who are demanding fairer pricing for journal subscriptions and increased open access from – you guessed it – Reed-Elsevier.[10] While I empathise with such protests (and have participated in them myself), I fear that they actually do little in the end to fix our damaged system. This is mainly because of the cunning way that publishing companies 'bundle' their subscriptions together, such that nearly all academic disciplines will have at least some of their most important journals bound into their library's subscription package. This means

---

[8] Apple was ranked first by Fortune 500 for profitability in 2015; the company had a 2015 revenue of US$182 billion, and profited US$39.5 billion.

[9] nodealnoreview.org.

[10] Reed-Elsevier is probably one of the worst offenders, but nearly all of the other big academic publishing companies have similar policies.

that for your average scientist, it becomes self-harming to boycott an entire company because you will miss the opportunity to publish in some of the most reputable and highly ranked journals in your discipline.

So, I think it is time to go back to the drawing board. It is a fair assessment that the big academic publishers benefiting financially from our hard work could probably afford to make the process a little bit fairer on us poor sods. What I am about to propose probably goes beyond what any single individual can do to change the culture, and I really cannot claim that the idea is entirely original. However, as scientific (and other academic) publishing continues to evolve, with more options and formats becoming available every year, I would be remiss not to attempt to get you thinking about this aspect when choosing where to publish.

These days we have a multitude of open-access models from companies such as the Public Library of Science (PLoS.org) that claim to be not-for-profit, academic-run models of review tender like Peerage of Science (peerageofscience.org), or even reviewer-recognition schemes like Publons (publons.com) (61). While these are, in my opinion, vast improvements[11] on the traditional pay-through-your-nose models that the for-profit publishing companies embrace, I still do not think they go far enough. But I am not going to discuss the relative merits of different open-access and other for-credit models; instead, could we not design a simple system of financial recognition for peer review? One could argue that any other model involving payment for services could jeopardise the quest for subjectivity reduction; imagine if we had to sell our science to the highest bidder – specialist-interest groups could conceivably highjack the system to their advantage. I would certainly give up the science baton if that ever happened.

But considering the frankly nauseating profits made by most publishing companies from our intellectual labour, I feel that there

---

[11] Open-access publishing is not all sweetness and light; there are many predatory publishers out there salivating to remove cash from your bank account in return for dodgy, poorly vetted publishing services. Beware.

is a strong justification for implementing a system of remuneration for our time, effort, and expertise without sacrificing our scientific integrity. Suppose for a moment that all peer-reviewed journals paid a standard sum for each review they commissioned. Let us say the price was something in the vicinity of €100 (the amount is somewhat inconsequential to the argumentation). Some back-of-the-envelope calculations suffice to demonstrate that this sort of payment is financially workable. I now receive in the vicinity of two to three review requests each week, and I refuse probably 80 per cent of them because of existing time commitments. So, I probably agree to do around twenty reviews per year. In a scheme where the journal requesting my review services paid me €100 per review, that means that I would earn an extra €2000 gross per year. I think you can agree that that does not represent a huge sum, but it is certainly enough to take my wife out to dinner a few more times than usual, perhaps buy some nice wine, or even (partially) finance an overseas trip for one of my PhD students. The point is that I would not be getting rich, but I would feel more validated and remunerated for my hard work.

What about the publishing company's bottom line? I will use my own field as an example; an average journal in ecology probably publishes between ten and twenty papers per issue, with 6–12 issues per year. I will therefore assume that this equates to an average of $15 \times 8 = 120$ papers per year. Of course, the journal would have to pay for the reviews of rejected papers too, so I will use a rather conservative 50 per cent rejection rate to conclude this exercise. At an average of two reviews per paper, that means $240 \times 2 \times €100 = €48,000$ paid to reviewers each year per journal. For the big scientific publishing companies who manage in the vicinity of 2500 journals (in fact as of January 2017, Elsevier managed 2958, Wiley 2410, and Springer more than 2500), this adds up to €120,000,000 (US$129,000,000) per year. That might seem like a big number, but when put into the perspective of turnover, it is only about 5 per cent of their annual profits. I therefore know that they could afford it.

Profiteering from slavery of any sort, including the intellectual kind, is not acceptable, so I feel perfectly justified in asking for such a modest proportion of profits for the work I now do for free. And there are other advantages to such a scheme beyond the improved economic fairness. As an editor, I would probably make sure I chose really good reviewers if the journal I was editing had to pay €100 for each review. As a reviewer, I would also most likely provide a much more in-depth, insightful review if I were getting paid to do it, and I would well do it in time (or else I might forfeit my fee). We could conceivably set up a review scoring system within each journal such that those reviewers who consistently provided high-quality reviews would get solicited to do them more often, thus gaining a little more than the lazier and less-dedicated types. Regardless, it is time for a change.

9   Constructive Editing

Perhaps it is just that I have been at this whole science thing for a while and I have become a bit jaded, or maybe it is a real trend. Regardless, many of my colleagues and I are now of the opinion that the quality of editing in scientific journals is on the downhill slide. Yes, all scientists complain about negative decisions from journals to which they have submitted their work. Being rejected is part of the process (Chapter 7). Aiming high is absolutely necessary for academic success, but when a negative decision is made on the basis of an appalling review, it is a little harder to swallow.

I suppose I can accept the inevitability of declining review quality for the simple reason that because there are now so many papers to review (Chapter 8), finding willing volunteers has become increasingly difficult. This means that there will always be people who only glance cursorily at the manuscript under consideration, miss the detail, and recommend rejection based on their own misunderstanding or bias. Obviously, it is far easier to skim a paper and try to find reasons to reject than actually putting in the time to appraise the work critically and fairly. This means that the traditional model of basing the decision to accept or reject on only two reviews is problematic because the probability of receiving poor reviews is apparently rising. For example, a certain undisclosed journal of unquestionably high quality for which I edit does not accept anything less than six reviewer recommendations per manuscript, and none that I am aware of is accepted or rejected based on only two reviews. But I think this is the exception rather than the rule – there are simply too many journals now of low to medium quality to be able to get that many reviewers to agree to review.

Being a great reviewer is not only part of being an effective scientist, it is also arguably a moral duty (Chapter 8). What goes around comes around; if you are a bad reviewer, you will receive bad reviews. But as you move up the career ladder, you will inevitably be asked to become part of a journal's editorial board and will most likely not be paid for the pleasure (Chapter 8). Even without remuneration, this is a noble thing to do in its own right, and it also gives you much better insight into the science-publishing process. While time-consuming, it is generally worth the effort, although I insert a warning here to the novice editors amongst you: becoming an editor is the fastest and most effective way to lose friends and make enemies. Although good editing etiquette is essential (more on that follows), even good editors become the frequent targets of author ire. Most reviewers to this day remain anonymous,[1] which is something I support, but editors themselves are almost always identified to the submitting authors during the course of the review. Thus, even the politest and most objective editor can suddenly find herself wearing a cowl of shit merely because she is the only identifiable human being in the process leading up to rejection. Editorships are therefore mixed blessings – they elicit recognition and respect from your peers, but they can also occasionally drive wedges between you and potential colleagues.

In terms of the act of editing itself, the first thing to remember is that you are not merely an administrator – you are, in essence, the overseer of scientific integrity. This is a huge responsibility that you cannot take lightly. The corollary is that if you fail to be a good editor, the entire scientific process also ultimately fails. This means, crazily enough, that editors actually have to *read* the manuscripts that they manage, in their entirety. It is not acceptable to skim the *Abstract* and make a decision based only on the reviewers' comments. As an editor, if you do not read the paper too, it is impossible to judge the quality of the reviews assessing it.

---

[1] While many journals encourage reviewers to identify themselves, at the very least to promote recognition (Chapter 7), more journals are moving to the 'double blind' model where both the authors and reviewers are anonymous.

Nearly all journals for which I have edited or am currently editing have sent annual e-mail summaries to the editorial board members that go something like this:

> The Journal of XXXX has had an x-fold increase in the number of submissions over the last y months. We therefore strongly encourage you to be extremely critical of what you let go to review. If a paper is not within the top $z\%$ of all articles in the journal, you should consider outright rejection instead of sending to review.

That sounds harsh, I know, but it is the sentiment behind the dubious 'space is at a premium in our journal, so we cannot accept all articles regardless of their merit' that you will receive as a primary excuse for most rejections (see Chapter 7). The main reasons to give such an excuse are to: (*i*) maximise profits for the publishing company, (*ii*) reduce the workload for largely volunteer editorial staff, and (*iii*) increase the journal's impact factor (see Chapter 7). It is up to you to judge the morality underlying this, but I do need to make it clear that editors are now increasingly directed from on high to reject papers as often as they can. It is therefore understandable that when a negative review comes in, the easiest thing to do is just accept it at face value and hit the 'reject' button.

But please resist that temptation. Instead, there are a few good ways to recognise a bad review, both the unfairly negative, or suspiciously supportive ones. First, if a review is only a paragraph long, be very, very suspicious of its quality and objectivity. Unless it is a review from someone I trust absolutely to give an educated and objective opinion, I nearly always dismiss it. Likewise, if the reviewer is clearly adversarial, or appears to take offence at the audacity of the authors even to write such rubbish, then take the review with a big shaker of salt. Next, it is a good idea to check the collaborative relationship between a potential reviewer and the authors of the manuscript you are handling – good mates tend to overlook major faults. Another tip is to avoid the 'lack of novelty' trap. Reviews that emphasise the former without at least attempting to critique the methods or results should perhaps be given a little less weight in your final decision. As I discussed in Chapter 7, 'novelty' is a poisoned chalice – do not succumb to the Luddite thinking that all papers have to be Earth-shattering. In fact, true novelty is extremely rare.

After deciding on whether a review should be believed in its entirety or dismissed as a poor-quality and subjective tirade, then the next thing to do is weigh the evidence for or against rejection. If you have one high-quality review (whether negative or positive) and one poor-quality review, I contend that you simply cannot make an informed decision. Of course, it is also necessary to add your own appraisal, but I would always recommend in such cases to obtain at least one more high-quality review before deciding which way to go. This will of course extend the time to publication, but to me this is far more important than rejecting a really good paper based on a shaky assessment, or worse, accepting one that is fundamentally flawed.

Although it sounds superfluous to mention, being an *editor* means that you have to *edit*, and I strongly urge you to do it well or not do it at all. It is an essential component of the greater process of scientific knowledge generation.

PART II The Numbers

10 **Fear Not the Numbers**

The title of this chapter might seem to be a bit odd to some of you, because science is an endeavour that absolutely must embrace numbers to function. As I argued in Chapter 8, science is a method to reduce subjectivity when answering questions via the careful measurement of phenomena, the control of confounding conditions, and replication to reduce the chance of falsely ascribing trends to random patterns. Measurement of any type yields numbers, and the discipline that has developed all the tools required to deal with numbers is mathematics. The results of these mathematical treatments must then be transformed by language into a coherent story, so that the results can be understood, repeated, and critiqued. Thus, the two *most important* skills a scientist must possess are *mathematics* and *writing*. The entire first part of this book is devoted to the latter, but I have not yet really discussed enough the former.

If you are a mathematician proper, or are perhaps in a field of theoretical physics based entirely on mathematical approaches, then all of this will seem almost patronising to you. I am really not speaking to you specifically; instead, I am making something of a plea to the other disciplines of science, from the biological, genetic, medical, chemical, geological, to the geographical, and so on – mathematics is and will continue to become more and more a necessary backbone to all scientific endeavours. The technological revolution alone, from next-generation sequencing in genetics, massive meta-analyses in medical research, global circulation models in climatology, to isotopic validation models in geochronology, requires not only increasing computing power and memory, but also ever-more complex and sophisticated mathematical frameworks to deal with the uberbytes of data these technologies generate. Mathematical prowess is therefore a prerequisite for all future scientists.

Yet paradoxically, mathematical training in many countries is on the wane, from public schools through to universities. In many parts of Europe, North America, Australasia, and I wager many other regions, up-and-coming science students have less and less primary training in the basics of mathematics. In fact, I also wager that many of you reading this might be proverbially nodding your head in reluctant agreement to my synopsis about your very own mathematical training and capabilities. But another group of you might still remain unconvinced that mathematical ability is central to your science interests, for if you are a clinician engaged in drug trials, a field ecologist collecting insects, or a forensic geneticist isolating DNA, then your skills and knowledge are primarily manifested in other ways. That might be true, but I argue that it is increasingly difficult to avoid mathematics in almost every discipline of science.

Increasing specialisation of the scientific disciplines means that if you are not necessarily inclined to maths yourself, you are probably finding that you are increasingly forced to collaborate with mathematical modellers, statisticians, or other similarly mathematically oriented people to analyse your data. This is, of course, a perfectly acceptable course of action, because no matter how intelligent or skilful you might be, there is always another scientist out there who will know more than you in a related and complementary field. Good collaboration is the key to success; however, if you and your collaborators speak entirely different (technical) languages, your shared projects are unlikely to delve more than superficially into the complexity of modern scientific questions.

All this preamble is merely to convince you that maths has a strong place in science, so if you find that your mathematical prowess is insufficient, then you might wish to do something to rectify your inadequacies. Thankfully, there are many ways to achieve this. One of the most effective ways to do so is to get more involved in the mathematical side of your very own data, and the simplest way to

achieve this is by working closely with a colleague who has at least a slightly better understanding of the maths than you do. Instead of merely firing off your data to the 'stats person' in a rambling e-mail, a little additional effort to spend some quality, one-on-one time with that person will almost certainly increase your base understanding of the specific techniques required. There is no better way to improve your own mathematical skills than being forced to treat your data yourself in the presence of a kind, helpful expert looking over your shoulder.

In fact, as a laboratory leader this opportunity is usually rather common in the form of your own lab membership. Although it might require a healthy swig of humility tea, sitting down with your post-doctoral fellow or even PhD student to understand how *they* are tackling the problems that you assigned to them can be remarkably instructive. I am no longer arrogant enough to deny the convenient truth that many of my staff are in fact far more switched on about how to analyse the lab's datasets than I am; after all, I hire or appoint them for this very reason! As the old saying goes, two heads are better than one (and preferably, dozens of heads are smarter still). The interaction can be as simple as a 30-minute session in your office going over the specifics of a particular statistical test, or it can be a more 'formal', workshop-like environment in which members of your lab present a problem and then collectively solve it together. I have also often asked a PhD student or postdoctoral fellow to present her or his current project as a lunch-time seminar or within the context of a lab meeting (see Chapter 13), focussing not so much on the results, but more on the analytical techniques employed. You will no doubt be surprised by how much you will learn this way.

Other ways that are much less interactive, but that can be equally rewarding, are to do a little online or textbook sleuthing. If you happen to require coding for any of your data manipulations or analysis (more on coding below), then most of the coding languages have online community forums where hordes of uber-geeks happily divulge the latest solutions to the most vexing of mathematical and coding conundrums. A little online searching will reveal this treasure trove of good advice, hints, and shortcuts. There are, of course, many good mathematical/analytical textbooks that are discipline-specific to which you can readily refer, although beware that some require a good deal of background knowledge to interpret easily.

It is also worth considering to attend the odd short course or analysis workshop when they are offered. It is in fact far too common that scientists become set in their mathematical ways and insist that all scientific lines of enquiry can be solved using the same techniques that they learned during their PhD. The reality is instead that techniques, just like datasets, evolve with time, so taking the opportunity to modernise your mathematical skill base is the sign of a dynamic and constantly evolving scientific mind. These sorts of short courses can take the form of anything from generic training sessions, to workshops dedicated to specific techniques. I have even been successful in convincing experts in specific techniques to come to my university to teach week-long courses to my lab members in particular, but also offering the opportunity to anyone else who might be available and interested.

On the subject of coding, I have known many a colleague whose face turns white at the mere thought of sitting in front of a blank console with a flashing cursor waiting with infinite patience for the first line of coded instruction. If you are the sort of person who needs a point-and-click GUI[1] environment like Microsoft® Excel or some other equivalent spreadsheet application to do even the simplest of data manipulations, then I can imagine the prospect of coding is equally frightening. However, off-the-shelf computer programs designed to handle any sort of statistical or other mathematical question in science are in fact relatively rare, mainly because the diversity of data types and complexities, and analytical requirements, mean that generic platforms are all-but-useless for most applications. Hence, the evolution of more intuitive and easy-to-learn coding languages and platforms has revolutionised how scientists analyse their datasets. Languages and platforms such as the R Programming Language,[2] Python,[3] Perl,[4] MATLAB,[5] AD Model Builder,[6] and many others, are becoming increasingly popular in nearly all branches of science to do the complex number crunching we need to remain competitive. It is my sincere recommendation to find a computer language and modelling platform that will satisfy the requirements of your research programme, and learn it. Likewise, appointing students and hiring technical or research staff familiar with different coding platforms is a great way to maximise the mathematical abilities of your lab and the research you produce.

I cannot reiterate enough that the essence of an effective and successful scientific career is the composition of your lab (Chapters 13–15), so making sure that your students and staff are adequately

---

[1] graphical user interface.
[2] www.r-project.org (free).
[3] www.python.org (free).
[4] www.perl.org (free).
[5] mathworks.com/products/matlab.html (requires purchase).
[6] www.admb-project.org (free).

trained in the mathematical treatment and manipulation of your datasets is paramount. Therefore, my advice is to make such skillsets integral to your selection process, as well as to their ongoing development during their time in your lab. A vibrant and mathematically adept lab membership will do nothing but enhance your scientific profile, even if you (mistakenly) view mathematics as a necessary evil rather than as a primary focus of your research.

Finally, I think that all scientists need to address the problem of waning mathematical (and scientific, in general) understanding amongst the non-scientist public. It is therefore our duty to address these problems head on, such as perhaps volunteering some of your precious time to visit a school or two in your neighbourhood to emphasise to the pupils just how important their basic mathematical skills will be in the competitive world in which they will eventually have to find meaningful employment. A forward-looking academic scientist with any sort of teaching responsibility will also insist that her students be flooded with mathematics throughout their education, from undergraduates to postgraduates, and beyond. If you happen to work in a university, insisting to your administrators that more mathematical training is required for the general studentry is the kind of pressure you can exert across your discipline and beyond.

11 Keeping Track of Your Data

> Good data management is not a goal in itself, but rather is the key conduit leading to knowledge discovery and innovation...

Mark D. Wilkinson and co-authors 2016 (62)

Despite good organisation and record-keeping being the pillars of efficient scientific enquiry, it surprises me just how disorganised most of us are. As a young scientist, I was particularly awful at organising my hard-won data, even for my own purposes, let alone for anyone else's. Back in the medieval days of 100-megabyte hard drives, floppy disks, and unreliable back-up systems, data management for most scientists consisted of scrawled lab- or field-books, poorly annotated spreadsheets, and reams of ASCII[1] text files. If we had to transfer data between computers, or Darwin forbid, between colleagues living in different cities or countries, the complications were manifold. If you ask scientists my age for the data they collected and analysed during their PhD, you might find that many, if not most, either cannot locate them, or even if they can, they are probably in an entirely inconvenient, unannotated, or irretrievable format. For example, what would you do now if someone handed you a box of floppy disks full of data and said 'Go for it'? As I described in Chapter 10, the data now being produced every second of every day make those dark, old days seem risible, yet in many ways we are in a worse position regarding how we store, manage, transfer, and protect our data simply due to the sheer volume of data dominating our daily lives. In my perennial quest to help you avoid the same mistakes I made during my career, which is indeed the entire point of this book, I have some advice for you on how you can at least minimise the probability of losing data, and in

---

[1] American Standard Code for Information Interchange: the numerical expression of a character in computer language.

the process hopefully enhance your scientific efficiency, effectiveness, and innovation as well.

If you are a scientist just starting out, you will be happy to learn that despite the information overload, you are living in a time in which it has never before been easier to manage, store, and disseminate your data. Therefore, developing a good culture of data management now will enhance the rest of your career. For one, most scientific journals now either provide the capacity for authors to upload their datasets directly as 'supplementary information' on the publisher's own site, or they require a link to an online database where the data can be freely accessed and downloaded. As I explained in Chapters 4 and 8, the phenomenon of the 'supplementary information' section is a mixed bag in terms of utility and fairness, but often it is a reasonable place to store at least smaller datasets if you are given the opportunity to do so. Although this is not my preferred venue for data

storage and access, it can represent sufficient impetus at least to clean up the dataset you used for that specific article, including 'metadata' that describe to anyone just what the different fields mean, what the units of measurement are, and perhaps any limitations or biases. This therefore represents an extra little kick in the bum to be slightly more organised than you would otherwise tend to be.

ONLINE DATA REPOSITORIES

A more long-term and robust solution to the problem of data management and availability is the increasing accessibility and sophistication of online data repositories that are specifically designed to store, organise, and annotate data from a wide variety of sources. For the article-specific datasets mentioned in the previous paragraph, these sorts of repositories are certainly my preferred option, mainly because they add another layer of scrutiny to your datasets that might not necessarily eventuate if left to you alone. For example, a metadata description that might seem perfectly clear to you (the author) could instead be utterly opaque to another scientist, so it is extremely useful to have a third party check to see that what you have written about 'column A' in a delimited text file is in fact obvious to anyone. Be warned though – uploading the same dataset many times because your descriptions are insufficient can become a little inconvenient and annoying, so try to see the bigger picture regarding the importance of doing so before abandoning the whole affair.

Repositories of this nature are not usually free of charge, however, because at the very least, you have to pay for the utility of having someone check that your data are in the right format. The fees also include payment for the servers and their upkeep, as well as any other custodial components of data management (salaries for database managers, replacement of hardware, online interface design, etc.). Sometimes you can be lucky enough to work in a field that has invested heavily in data management, so some discipline-specific databases might be free for the average user. In my own field (ecology and evolutionary biology), there are several such repositories available to my community of researchers, including the Dryad Digital Repository[2] based in the USA, the Advanced Ecological and Knowledge Observation System[3] (ÆKOS) based in Australia, DataBasin[4] from the Conservation Biology Institute in the USA, and the Global Biodiversity Information Facility[5] in Denmark. However, this is only a tiny slice of the wonderful array of general and discipline-specific online data repositories available to scientists today. In fact, a quick perusal of the Open Access Directory[6] or the Nature Publishing Group's recommended data repositories for the journal *Scientific Data*[7] will lead you to discover over 100 repositories in fields as diverse as chemistry, physics, medicine, geosciences, geology, energy, ecology, computer science, archaeology, and astronomy.

A simple rule of thumb to follow here is that upon nearing the completion of a manuscript, you should start thinking about creating an easy-to-understand and adequately described (via metadata) dataset that you can submit to an appropriate repository. I tend to do this before I submit the manuscript to a journal, because it can take some time to get everything in the right order and obtain a digital object identifier[8] (DOI) for the dataset itself. Once you obtain your dataset's DOI, you reference the link in your manuscript and register it with the journal's submission system, and job done.

While there is an entire field of information science dedicated to data management, I suspect that most of you will end up submitting something akin to a basic spreadsheet. While this is acceptable, it pays to know that databasing can be much more sophisticated than this. It is really up to you how far you want to take things. If you are not the type of person who relishes developing the necessary practical

---

[2] datadryad.org (charges a fee).
[3] aekos.org.au (free).
[4] databasin.org (free up to 1 GB).
[5] gbif.org (free).
[6] oad.simmons.edu.
[7] www.nature.com/sdata/policies/repositories.
[8] A unique alpha-numeric code to identify published content and its internet location (doi.org).

knowledge to become a database manager, there is nothing to worry about. Even the average database Luddite now has access to excellent guidelines for making sure that their data are as useful as possible to the general scientific community, and that they can be found, accessed, and queried both manually and automatically (62). While data management is not always the most exciting or captivating part of being a scientist, a little bit of planning and foresight will ultimately free up more of your time for doing the more pleasing tasks involved in an effective scientific career.

DO YOU KNOW WHERE YOUR CODE IS?

Just as access to well-described, organised, clean, and intuitive datasets is important, so too is good management of your computer code. Even if you are not the hard-core, computer-science type, chances are that whatever your scientific discipline, you, members of your lab, or at least some of your students will end up generating code in some computer language. I admit that I am not the archetype of good code management myself, for my code is typically strewn across my computer's file structure, and I often cannot easily find bits of code that I know I had successfully written to solve particular problems in the past, and am now facing the same problems for a different analysis. While I have improved the organisation of my code over the years, I urge you to save yourself the heartache and develop a good code-storage framework and intuitive version control from the beginning.

Fortunately, there are many great online tools designed for just that sort of thing. GitHub.com is probably the world's largest repository of computer code, providing a version-control platform designed particularly for teams of developers to collaborate on designing software applications in a controlled, structured, and open manner. If you are already a serious coder, then you probably already use GitHub, or some other analogous platform (e.g., SourceForge.net, Launchpad .net). If you are more of a code-dabbler like most scientists, then

GitHub is perhaps something you might consider using to organise, update, share, and archive your precious code. At the very least, GitHub allows you to keep track of your different code applications in one handy place, and it lets you share them with your colleagues and students. GitHub not only acts as a publicly accessible repository for code, it also can store and share relevant data and documents. This means that you can refer to your code profile and the links to specific projects in manuscripts that you submit for publication.

TO SHARE, OR NOT TO SHARE, IS NO LONGER THE QUESTION

Throughout this chapter I have been tacitly assuming that you are indeed entirely fine with the idea of having your hard-won data spread across the internet, and that anyone can access and use them. In reality, many of you are probably not comfortable at all with that concept, and that the very notion of 'sharing' your data with anyone but your closest and most-trusted colleagues is the stuff of nightmares.

I too was once far too concerned about the privacy of the data for which I had literally sweated and bled, for I feared that some nefarious and amoral scientist would steal, analyse, and publish them before I had the chance, thus usurping my unique contributions to the body of human knowledge; in other words, I was worried I would not receive the accolades and recognition for my work. Perhaps I was just a little more paranoid than your average person, although I still encounter such attitudes today. While data theft can occur, in reality it is unlikely in the extreme that anyone would bother trying to out-do you in this regard, mainly for the simple reason that in most cases, data availability is not normally the limiting factor for scientific advancement. Another reason why this should not worry you is that far too few of us have the time to publish all of our own data, let alone someone else's.

As it turns out, *not* sharing your data can in fact reduce your publication opportunities (and thus your scientific reputation) because your colleagues are less likely to know what you have been doing. By all means, it is wholly reasonable to keep the specific data upon which a particular manuscript is being constructed under lock and key until you submit the resulting manuscript to a journal, but this is normally within a timeframe of months rather than years or decades. But once you have started to publish from a particular dataset, then my sincere advice is to open it up to all and sundry.

Why? Even if you have the intention of mining your dataset for more analyses and stacks of new manuscripts over the coming years, making it available to the greater research community is more likely to make new opportunities rather than stealing them away from you. The most basic reason that this is true is that if some other research group ends up using your data, they will at least be obliged to cite you in some form, and as I outlined in Chapter 6, getting citations is good for your career. Another advantage of this type of second-hand sharing is that more often than not, people using your data generally want to understand how you collected them, what limitations they might have, and if there are subtleties not necessarily covered in full in the accompanying metadata. To do this, they will generally try to contact you directly, ask some pertinent questions, and in many cases, invite you to join their authorship team. Yet another benefit of data sharing is that different people have different perspectives, analytical techniques, and interpretations, such that alternative applications can identify new lines of inquiry and allow you to pose and test hypotheses that might have otherwise escaped your attention. These days, there are even journals devoted entirely to database descriptions (e.g., *Scientific Data, GigaScience, Standards in Genomic Sciences,* etc.), so the database itself represents a publication opportunity. In short, data sharing has so many more pros than cons that remaining a data hoarder is patently to your disadvantage.

As I mentioned previously, many – if not most – scientific journals today do not really give you the option any more to be a data miser. In my own field, it is becoming rare for a journal to agree to publish your manuscript if it is not at least accompanied by an easily accessible dataset, which more and more commonly today means a link to a database in an online repository and an associated digital object identifier (DOI). In other words, you tend to have diminishing opportunities to choose to limit the extent to which people can access your data, meaning that ultimately, it is really no longer up to you. Regardless, some scientists still attempt to limit access to their online data by not including all the necessary information or metadata needed to interpret them correctly – this is simply childish and self-harming in the long run. Instead, I strongly urge you to make your online datasets clear, complete, and well-described (e.g., via intuitive metadata). Your colleagues will not only be grateful, they will be more likely to do the same and share their data and publication opportunities with you.

BE WISE ABOUT WHAT YOU PUT ONLINE

There is one more aspect of making your data publicly available that I have not yet discussed, and that is the prickly issue of whether your data contain sensitive information. Of course, there are many different types of 'sensitive' information that might accompany the more basic quantitative measurements of your datasets, with perhaps the most common being personal details of any human subjects. For example, if you are a medical researcher and your data are derived primarily from living human beings undergoing some procedure, trial, or intervention, then clearly you are bound by your human ethics approvals *not* to publish information like names, addresses, or anything that could be used to identify the subjects in your sample. In fact, human ethics approvals generally prohibit any sort of public accessibility to medical data that has personal information included; thus, the scientists concerned are being pulled in two different directions – keeping their subjects' personal information out of the hands of the public, while still making the data available to other scientists.

There are ways around this, such as publishing only generic information online (i.e., by excluding personal identifiers) that could then be linked to the more sensitive data via unique identifiers. In these cases, any other researcher requiring the additional information would have to seek specific permission from the primary researchers, pending additional human-ethics approvals. Survey data, where individual people are asked questions on anything from their personal habits to their voting preferences, can also come under this umbrella of data sensitivity. Even data mined from social-media platforms can be restricted on privacy grounds. Often commercially sensitive data are in the same category, such as mining leases, fishing grounds, and hunting sites. It is therefore the responsibility of both the researcher and the committees granting ethics or permitting approvals to decide on an optimal trade-off between data dissemination and the protection of individual privacy or commercial privilege.

Many other types of data sensitivities abound, such as those that must be considered given the propensity of certain nefarious types to exploit scientific information for their own personal gain. One rather galling, generic example of this comes from the fields of archaeology and palaeontology, whereby scientists who have published the location of deposit-rich sites have been horrified to discover that curio and fossil hunters have pilfered precious specimens after reading the relevant scientific articles. As a result, most online archaeological or palaeontological datasets today either do not publish the site locations at all, or they deliberately add a location error so that only specialists will know where to look (63). An even more disturbing behaviour is becoming increasingly frequent as would-be poachers and pet-traders use the scientific literature in ecology and biodiversity conservation to discover the locations of rare and endangered wildlife and plants (64). By virtue of being rare, many species are considered highly valuable in the trade of parts or pets – just think of rhino horn, rare orchids, and tropical aquarium fish. If you happen to research any rare species, fossils, human remains, or other potentially valuable specimens, do think carefully about what data you make publicly available online. At the very least, do not tell people where you found them.

12 **Money**

I will begin this chapter with a *proviso*: I am not the most financially organised scientist in the world. In fact, I dare say that I have rated myself amongst some of the worst in terms of balancing budgets, knowing the current state of my research funds, and planning ahead for my research's financial requirements. But I suspect that I am far from alone in this regard amongst most scientists, for I would most likely have gone into accounting or business if I had a good head for finance. But I do not, nor probably do many of you.

To recall my sincere desire expressed in the *Preface* about having had the opportunity to foresee what skills I would need to acquire to be an effective scientist, I now lament that my lack of accounting skills and misunderstanding of the funding world probably eroded my capacity to do even more good science. But as they say, 'wish in one, shit in the other, and see which one comes true first'. Perhaps my clumsily acquired wisdom can help you to become an effective scientist flush with cash to fund your every research desire.

WHERE TO SEEK FUNDING

For most scientists working outside of medical research, there are usually only a few agencies in any one country that will offer you funding opportunities. I have singled out medical research here because human beings are terribly concerned about their own well-being, so understandably we as a society have decided to invest heavily in our own medical welfare. This means there are many different public, private, and corporate funds available to finance medical research. I am not complaining,[1] but merely attesting to the general principle that

---

[1] Well, maybe just a little.

scientific funding agencies are generally not as ubiquitous (or generous) for other major science disciplines.

I also mentioned 'country' above because scientists are often limited to apply for funds within their own nation. This funding bottleneck arises because research priorities are typically set for the benefit of people within the funding agency's national jurisdiction. This is not to insinuate that there are no international funding agencies – there are of course many. But unless your research is tied to foreign aid (whether medical, agricultural, hydrological, or otherwise), or is specifically designed to align with an international funding agency's brief,[2] these can be a little more difficult to access than most national funding sources. If you happen to be doing research in an area where international funds are accessible, then by all means, do your utmost to apply for them. Regrettably for the rest of us, we are normally restricted to our taxpayer-funded, government science-funding agencies, such as the United Kingdom's *Medical Research Council* and *Natural Environment Research Council*, New Zealand's *Marsden Fund* and *Health Research Council of New Zealand*, the USA's *National Science Foundation* and *National Institutes of Health*, Australia's *Australian Research Council* and *National Health and Medical Research Council*, Canada's *Natural Sciences and Engineering Research Council of Canada* and *Canadian Institutes of Health Research*, Finland's *Academy of Finland*, and so on.

In most national contexts, academic scientists gain additional esteem just by being successful at attracting these government sources of research funding. While the amount obtained might in many cases be less than some non-government sources, the additional esteem attached to national-agency grants cannot be understated. Regardless, I have yet to meet a university administrator who does not encourage the academics they manage[3] to have as much funding income as possible, irrespective of the source. In many cases too, such

---

[2] European Union (EU) grants are a good example of this, but only if you are a citizen of a member state.
[3] persecute.

government research funds are restricted to academic scientists (i.e., corporate or government scientists are not eligible to apply), so one must ascertain eligibility in each particular case. If you happen to be a government or corporate scientist reading this (despite the book's subtitle), then the issue of funding esteem is less relevant, and particularly in the corporate world, funding tends to become the concern of administrators anyway.

My advice to this point has mainly pertained to scientists in 'developed' nations, but it starts to lose relevance for scientists from nations still part of the way along the development pathway. Poorer countries, or those with unstable governance systems, are notorious for having few, if any, funding opportunities at the level of national government. African scientists in particular are chronically underfunded, even in terms of salary (65), and they even have fewer international funding opportunities. This has led to many young African scientists abandoning their home soil for greener funding pastures in developed nations (66, 67), with most never returning (66, 68). While there is no easy solution to this dilemma, *ex situ* appointments from developed-nation institutions could potentially help; otherwise, my only other advice would be to collaborate as much as possible with foreign colleagues who themselves might have more funding opportunities.

I am deliberately writing in general terms regarding funding sources because every discipline has different opportunities, and each national agency imposes different eligibility criteria. Crowdfunding is also perhaps one way to obtain research money that is only just starting to take off in certain scientific circles; if you have a great project with a cool story behind it, can obtain the support of colleagues and friends, are outgoing and can 'sell' your pitch, and are willing to make compromises, crowdfunding[4] might be something you can investigate (69). Another source is philanthropic donations,

---

[4] Many online crowdfunding platforms are available to organise and campaign your projects, including *inter alia* pozible.com, gofundme.com, and kickstarter.com, as well as some science-specific ones including experiment.com, crowd.science, petridish.org, and consano.org.

but the rare opportunities here are also highly institution-, discipline-, and country-specific (70). Scientific research is unequivocally expensive, so do not expect your employer to hand out gratuitous buckets of cash whenever you need new laboratory equipment, reagents, field vehicles, or salaries for your staff. Obtaining funding is a preoccupation that will last your entire career, so the better and earlier prepared you are to master the art of convincing people to give you money, the more effective a scientist you can become.

HOW TO WRITE A GRANT PROPOSAL

The title of this section might promise a lot, but it would be disingenuous of me to imply that I could cover all of the essential components of this massive topic in one chapter. Many people have made careers out of teaching people how to write successful grant proposals,[5] so I will not pretend to be comprehensive and insult their expertise. That said, I have been reasonably successful on the grants' side of the science game, and I have assessed a fair few grant proposals over my career. I therefore contend that I can offer you at least a few major pointers. As usual, each person probably has her or his own way of doing things, so there is unlikely to be a single, winning method for writing all grants across all disciplines of science. Approaches will also vary by funding agency and country of origin. I am therefore writing this chapter primarily for earlier-career scientists who have yet to become fully indoctrinated into the funding cycle, with generalities that should apply to most grant proposals.

1 A Proposal Is Not a Scientific Article

In the long list of things they never taught you as a student, but need to know to be a successful scientist (i.e., this book), this has got to be one of the most important. Grant proposals cannot and should not follow the standard format of peer-reviewed articles. Articles tend to put an elaborate background up front in an introduction, then a complex description of hypotheses, followed by an even more complex description of methods and results. Do not do this for a proposal. A proposal should be viewed more as a 'pitch' that hooks the assessors' attention from the moment they start reading it.

2 Understand What the Funder Actually Funds

Many neophytes to the eternal funding cycle have the mistaken assumption that funding agencies exist to fund the sort of research that the researcher wants to do. I wish it were so. However, a funding agency exists to fund the type of research that the *funding agency* wants to see happen. It does not matter if it is the Australian Research Council, the Gates Foundation, the Wellcome Trust, the National Science Foundation, or the Slovenian Science Foundation – they all have

---

[5] In fact, my wife is a career grant-writing consultant and trainer.

(often government-set) funding priorities. It is the applicant's responsibility to know what these are and follow them. If you must alter
your research desires to fit within these frameworks, then so be it,
and avoid being offended that no one wants to fund what you want to
do.

3   Read the Guidelines and Follow Them

Again, this might seem bleeding obvious, but few people actually
read the guidelines before launching into the draft proposal. I recommend that no matter how boring the guidelines are (and they are
always mind-numbingly boring), avoid this mistake. I also recommend spending days, if not weeks, examining each element of the
guidelines that are provided to tell you more than just the margin
or font size, which the members of the funding agency's assessment
panel expect. There are many subtle hints in the guidelines that tell
the applicant what and how to write. More importantly, they are usually rather explicit about what the applicant *should not* include. This
is in fact far more important than generic content guidelines because
it could mean the difference between normal assessment and disqual-
ification (i.e., the proposal on which you spent months could be sent
to the bin without even being assessed – a dreadful shame).

4   Assume the Assessor has No Knowledge Whatsoever of Your Field

If you were to write a proposal to your immediate supervisor, closest colleague, or a specialist reviewer, you would assume that your
audience has an adequate background against which they can judge
the merits of your work. Most assessors will not have this experience or knowledge because of the increasing specialisation in science
I mentioned earlier. This means that you should write your proposal
with a not-quite patronising tone that appears, at least to you, to over-
explain even basic concepts in your particular area. A general rule of
thumb therefore is not to assume the assessors will have any existing
knowledge, so you must define all your terminology in the simplest

of terms, and repeat *ad nauseum*[6] why the subject is important and compelling.

5 You Are Selling Yourself As Much, If Not More, than the Research Project

A pessimist might justifiably conclude that as long as the proposed research is not flawed and fits the funding guidelines in question, many agencies could not give a rat's hirsute testicles about the subject matter per se. Instead, funders focus mainly on the reputation and track record of the person(s) proposing the research. Early-career researchers could then (again, justifiably) conclude that the system is stacked against them because if they do not yet have a great track record, what chance would they ever have of getting funded in the first place to build one? *Voilà* one of the many deep problems with the standard funding models in science. It is a truism in society that the rich get richer, but paupers (i.e., early-career scientists) can avert this to some extent by careful alignment with more established colleagues and by placing an emphasis on what makes them exceptional.

The take-home message is that *you* matter so very much in the proposal, so you must sell yourself and your team of investigators (if you are more than one). After all, a research grant is a type of award, so think of the proposal as an application for a prize. Attempt to make your assessors think that you are the greatest thing since $E = mc^2$, and focus on your career highlights and contributions. The funding agency and their assessors need to know that they will be giving money to a person whose proven reputation, skills, and output will guarantee the return on their investment (i.e., the success of your proposed research). If you are the shy, introverted type, it is really a good idea to ask a more outgoing colleague to ascertain whether you have undersold yourself. It might seem a tad conceited and perhaps braggadocious when you read aloud the section describing the team's

---

[6] I would not recommend trying to make your assessors literally sick by your repetition, but by the time you submit your proposal, you might be ready to vomit.

track record, but this is an essential marketing strategy for successful grant applications. If you do not make yourself sound great, your competitors will assuredly make themselves sound better.

6 Never Underestimate the Value of Good Collaborators

This might now seem obvious from the previous point, but I am not referring here to the parasitic tendency to latch onto the best person in your field to maximise the probability of success vicariously. A careful selection of high-quality, proven researchers that act to fill any weaknesses or gaps in your expertise, reputation, and knowledge will make a colossal difference to how your team's capability and the project's feasibility are assessed. In contrast, it is equally important to exclude weak collaborators who, at least from the perspective of the assessor, bring little to the team's expertise or reputation. If the assessors are convinced that the investigator team could realise the proposed research better than some other mix of scientists, then you immediately put yourself in the running for funding.

7 Never Underestimate the Value of a Good Title

In reference to the point above about 'selling' yourself, a catchy title and a clever opening will potentially get you the attention you need to stand out amongst the hundreds or thousands of other researchers vying for the same pot of funding gold. Some tips about good titles include: (*i*) avoiding question formats (it is not going to make people inherently more curious – after all, you will not have the results yet to answer the question), (*ii*) using terms that are immediately understood by a large range of non-specialists, (*iii*) being short and to the point, and (*iv*) containing something to tell the assessor why the proposed project is so bloody important that you should be given oodles of cash.

8 You have Already Won or Lost in the First Page

Including the title, an assessor will at the very least get bored, or abandon any further assessment at worst, if you have not explained on the very first page (*i*) *what* you are intending to do in specific terms, (*ii*) *why* it is so exciting, and (*iii*) *which* major societal problem it will solve (more on applied versus theoretical subjects below; see also Chapter 23). You have to keep the assessor's attention because she will be reading sometimes hundreds of similar proposals, so it is in your best interest to provide some tantalising information up front that will encourage the assessor to keep reading. Avoid the temptation of couching the *what* aspect in generic or motherhood terms such as 'breaking ground', 'improving knowledge', 'applying novel techniques', even though you will have space to provide details of the approach later in the proposal. As an assessor myself, I find it rather frustrating[7] when I must read eight pages of prattling text before I have even the faintest idea of what the investigators propose to do (cf. achieve). An assessor wants to see immediately that you will be proposing to do x, y, and z such that you can answer big questions a, b, and c efficiently, effectively, and convincingly.

9   Describe Why Your Proposed Research Is Exciting

You might think your research agenda is exciting, so might your partner, and your grandmother, but these are not guarantees that anyone else does (yet). Do not take it as given that your chosen topic (notwithstanding Point 2 about aiming your research plan at the funding agency's targets) is of any interest whatsoever to anyone else. You have to explain and convince, in gruesome detail, what makes it so fascinating and essential that you should do the research now. As an assessor, I want to end up being as excited by the proposed project as you are, so your sell has to go beyond the mere intellectual appeal and invoke an emotional response as well.

10   Explain the Applied Outcomes of the Research

I am not going to debate the relative merits of so-called *blue skies* and *applied* research here (I think they are both necessary – it is really only the ideal ratio that should be debated; see also Chapter 23), but there is

---

[7] You should probably avoid doing anything that will frustrate an assessor.

a reasonable chance that if you have followed Point 2 and aligned your project well with the agency's *raison d'être*, there will be some application to your work. In other words, you should ask yourself throughout the application-writing process why society should invest in your research if nothing practical would ever result. I recommend spending a good deal of time explaining how your eventual results will improve something for at least someone somewhere, either through policy, technology, or remediation. Never proclaim that your research will simply provide humanity with more 'knowledge' and not much else.

11 Funding Agencies Are Generally Risk-Averse, So Make Sure That You Have Some Relevant History

Coming back to Point 5, you must understand that most granting agencies are not willing to take a punt on your potential as a researcher, no matter how wonderful you truly are; instead, they want a guarantee that you will be able to do what you say you are going to do. Remember, you are competing with many others for a paltry sum. Proposing a difficult, elaborate, and risky research project will only lead to disappointment. It might sound a little jaded on my part, but it is true to some extent that funding agencies only fund what has already been proven to work. Sadly, if your project is too innovative (i.e., 'risky', as seen through the eyes of the assessor), it is unlikely that you will receive funding. A working rule of thumb is that if you have some established track record in the area of proposed research (e.g., a previously published paper in the subject), then you have a much better chance of success than proposing something that you have never done before.

12 State Your Hypotheses and How You Will Test Them

It still surprises me how few scientists know what hypotheses are, even those who have been at it for many years. I admit that I struggled with the exact meaning of *hypothesis* back when I was a postgraduate student, probably because my mentors did not have a clear understanding either. Hypotheses are the essence of the scientific method, so knowing that they are *testable* assertions and not merely *aims* will distinguish the crisp, clear research proposals from the vague, ambiguous ones. For example, it is one thing to aim to solve the energy crisis, but it is quite another to say how you will do it. Take care to list your main hypotheses and their predictions, and how you will test them with data and/or models. If you do not propose testable hypotheses, your risk of failure (as deemed by the assessor) increases, and so too does your probability of not being funded.

13   Avoid Jargon

At risk of sounding like a broken record, jargon is for specialists only. In my opinion, it should be avoided even in specialist articles because it inevitably retards comprehension and limits readership. It is even more important to avoid jargon of all types (and I include abbreviations, initialisms, and acronyms in this list) in a research proposal. If the assessor does not know *immediately* what you mean when you invoke a particular term, even after defining it earlier in the proposal, you will lose her attention. Be clear, even if it requires more words, and explain everything simply.

14   Quantify

A *motherhood statement* is defined as a vague, 'feel-good' platitude with which few would inherently disagree. In science, it is usually associated with some outcome (e.g., 'we must preserve species', 'we should reduce human suffering'). Why 'must' or 'should' we do this thing? Do not assume that your assessor has the same values as you, or that the funding agency's moral compass points in the same direction. Likewise, avoid subjective qualifiers like 'a lot', 'a multitude', 'very', and 'significant' if they have no quantifiable meaning. Make sure you demonstrate that the research will quantify a phenomenon or process and that you do not inadvertently demonstrate your biases or lack of understanding by using such subjective terminology.

15 Be Methodologically Specific

Some people try to hide the dirt under the rug in a proposal by including throw-away lines regarding how they will achieve their objectives. One in particular that I see far too often is '... and then we will model the system'. How will you model the system? What model(s) will you construct or apply? How will you parameterise these models? Do you have the necessary expertise in your team to construct such models? Likewise, '... we will measure...' and '... we will construct...' statements without the corresponding methodological detail (i.e., exactly how you will achieve them) are a clear demonstration to the assessor that you do not know what you are doing. If you do not really understand how, make damn sure to include someone in your team of investigators who does, and then describe their approach in detail.

16 Be Realistic

Proposals can often verge on the fantastical because the proponents purport to solve the mysteries of life, the universe, and everything in ten or fewer pages. Without lessening the impact of why the research is important, do not venture too far into Fairyland and claim that you will be able to solve all elements of the problem under investigation. Stick to the hypotheses and do what you can within the budget and timeline proposed.

17 Give Some Serious Attention to Your Communication Strategy

Many, if not all, funding agencies want to know how you will make the results of your research known to the public, and not just to the funding agency itself or to the few specialists who might actually read the ensuing scientific articles. Communication is becoming more important these days as society in general becomes more adversarial to scientific results and expertise (71). While social media might not be everyone's cup of tea (but see Chapter 21), spend a little more time explaining how you will reach a much broader range of people than most researchers achieve. Think of clever and innovative ways of reaching out, and dedicate more than a passing thought to this section of the proposal.

18   *Have an Experienced Colleague Read the Proposal*

After it is all said and done, give your proposal to someone with more experience than you to read and critique. Ideally, you will do the same with two or three other comparable colleagues. In many ways, this is the most important part of the process before you even submit to the funding agency itself. More opinions are better than one. Be advised, however, that you will be asked to do the same thing for your colleagues at a later date.

19   *Ask Whether a Layperson Would Fund Your Research*

Most research these days is publicly funded, so asking a few lay taxpayers that you can arm-wrestle to read the proposal will tell you if it strikes a chord, or bores them to suicide. I generally try to pin down a family member first if I can, because they often find it more difficult to say 'no' to you. This differs from the previous point because it involves a non-specialist. If you cannot convince the punter in the street, it is less likely that the funding agency will deem your research worthwhile.

I will close this section with a comment about *marketing* that typically baffles new scientists. While the quality of your proposed research must be immutable, nearly every other element of a good proposal comes down to good marketing skills. Scientists are not inherently good marketers because they do not tend to be educated in that profession, nor do they normally have much interest in the topic. In fact, most scientists I know unreservedly despise marketing because it symbolises the antithesis of the scientific method (i.e., facts are irrelevant; marketing manipulates emotions and subjective beliefs to elicit irrational consumer behaviour). Regardless, every effective scientist is also a good salesperson because she can convince others to give her money. The art of good proposal writing is therefore grounded in effective salesmanship.

KEEPING TRACK OF YOUR CASH

As I stated at the beginning of this chapter, I am certainly not a shining example of good grant management. I was typically bewildered by my students asking me how much of their departmental allocations remained, or by my postdoctoral fellows asking me if the grant from which their salary was drawn could spare some money to send them to a conference in Switzerland. My response was most often to send them to query one of the professional staff in charge of university finances. This approach is not an efficient or particularly intelligent way to manage your research funds, so with the benefit of some painful hindsight to guide me, I have a few pointers that you might like to consider once the grant money starts to flow in.

If you are an inherently organised person, then this section might not serve you well; however, if you are more the disorganised type like me, then there are a few things to consider when managing your funds. While nearly all grant proposals require the applicant to stipulate a budget, these often range from a basic 'guess' of what things might cost, to a detailed, sub-annual breakdown of itemised costs for all proposed equipment, personnel, and travel. A critical mistake is to neglect updating and expanding that budget once you know what you will receive. Even the best-planned budgets will have to be updated, either because you will not receive all the funds you originally requested, or because elements of the research inevitably change – such as anything like experiments not working, equipment breaking, salary rates increasing, or inaccurate initial estimates. When you are faced with less money than you had hoped to receive,[8] then a good idea is to have a back-up budget that assumes a substantially smaller total from which to draw funds. Whatever the outcome, I recommend spending a few days restructuring the budget

---

[8] From experience, I typically expect to receive anything from 60 to 80 per cent of what I asked for initially, no matter how realistic the budget. Any more should be considered a bonus.

once you understand how much, and at what frequency, your funds will be received. Add as much detail as possible, and try to avoid choosing the cheapest-possible option because otherwise, you will tend to underestimate expenditure over the lifetime of the grant.

These days, most tertiary institutions employ personnel specifically for organising their researchers' funds, and they typically operate through software packages that are designed to simplify the expenditure and acquittal of grant monies. Despite generally being overly complex, clunky, and non-intuitive, it is a good idea to get up to speed with the software package your institution uses so that you can acquit your expenditures as efficiently as possible. Such software also gives you close to real-time estimates of how much money you have left in the account, and what anticipated expenditures remain. Some software packages are better than others, but there is nearly always some sort of accounting software, and you will need to know how it works. It can also be helpful to transfer total balances to a personal spreadsheet on your computer or notebook in your lab so that you at least can keep a quick mental tally of where your finances stand at any given moment.

Another situation to avoid is to find out that you have money left in a grant account that you should have already spent. This often happens as you approach the end of the grant's lifetime and have been lax with your knowledge of associated expenditures and payment schedules. If you do find that you have money left, most granting agencies require that you justify why you have not spent all the money you requested, and typically they will ask for you to pay back the remainder to them. I am not fond of giving money back, so sometimes I can get out of the situation by impulse-buying some associated equipment, extending someone's contract for a few months, or applying for a 'carry over' to the new financial year. None of these actions is ideal, so avoid the problem altogether by becoming a better accountant of your own money.

One final recommendation pertains to reporting grant progress and completion. Often referred to as 'acquittal', this uncomfortable task is not only a good idea, it is usually required by the granting agency. If you are required to provide regular reports updating your progress, be honest about any setbacks or failures, and spend the time to justify your purchases. Final reports can represent good reality checks on your overall success because they remind you of your lack of progress, failures, inefficiency, and poor organisation. To circumvent these feelings of guilt and inadequacy, try to establish a good financial and progress-reporting mechanism done at regular stages throughout a grant's lifetime. Of course, you can ask your administrative counterparts to assist, but do not rely on them to keep you organised – only you can do that. That said, it is a wise practice to treat your institution's finance staff with respect and kindness – while they might stuff up from time to time, having friends who can help you out of a financial bind with your budgets is an effective way to avoid stress and disruption of your research.

PART III Good Lab Practice

13　Running a Lab

Running a laboratory can mean many different things to many different scientists. It can simply refer to a single researcher with a few PhD students, a technician, and perhaps a postdoctoral fellow or two, or a large centre of many mid- to high-level academics with 'sub-labs' of their own. There is no 'ideal' lab size, but unsurprisingly, larger labs tend to produce more papers (72), albeit not necessarily in the highest-impact journals. Other evidence suggests that a lab size of ten to twenty-five people reaches maximum efficiency (73), and I can attest that managing anything more than twenty people by yourself becomes a bit of a nightmare. Most people will end up managing labs somewhere between these extremes of magnitude, so irrespective of size (which varies temporally anyway), it is important that the group is managed well to maximise harmony and productivity. This sounds easy enough, but most scientists are not trained managers, and often their 'people' skills are somewhat deficient. A common perception, whether supported by legitimate diagnosis or not, is that many scientists have at least one personality disorder, whether that be some form of mild autism, narcissism, or Asperger syndrome.[1] It could be that science just attracts such people, or that our relative freedom means that we do not receive too much negative feedback for inappropriate displays of social awkwardness. Regardless, social ineptitude is not a great recipe for running a lab effectively.

Short of seeking professional psychological treatment for clear cases of borderline personality disorders, there are many things one can do to assist in the day-to-day process of managing a group of

---

[1] A type of autism in which the afflicted tend to be socially awkward and have an all-absorbing interest in specific topics. Sound familiar?

scientists. One thing I have learned is that scientists like to feel like they are part of something bigger than themselves, whether they be fledgling researchers completing a doctoral dissertation, or seasoned, globally recognised researchers. This requirement to 'belong' means that the lab head must endeavour to engender a culture of community, and the simplest way to do this is to schedule regular lab meetings.

Lab meetings, in many ways, are the primary social interactions for members of almost all labs, and they might even be the only times that all members can get together to interact. I must make this clear – a lab meeting represents much more than just an opportunity to go through standard house-keeping rules; it is in essence a social-bonding exercise that acts to develop a collaborative and interactive research environment beyond the mere exchange of empirical information. Depriving human beings – even scientists – of this essential social opportunity will inevitably breed discontent, and ultimately erode your lab's overall productivity.

There are, of course, many different models for running lab meetings, but they all include most of the following. First, they tend to be scheduled at regular times and days – some like the entire lab to meet at least once weekly (usually, smaller labs), while others schedule fortnightly or even monthly get-togethers. It will normally be impossible for all members to find an ideal time and day to be together, but with a little discussion, a time conducive to most people's schedules can usually be found. I recommend at least a fortnightly schedule where the day and time remain fixed for at least most of the year. New cohorts of researchers and students, or new academic years, can be opportune moments for renegotiating the schedule if they cease to be ideal for most lab members. Another good idea is to schedule them during meal or break times, or even to incorporate another social lubricant such as Friday drinks. Everyone needs regular breaks anyway, so if you can encourage your team to bring along a lunch, a coffee, or even a bottle of wine (circumstances allowing), then you can kill two scheduling birds with one stone, so to speak. Resist the temptation to have regular meetings at the pub though, because the conversation will quickly dissolve into more mundane, decidedly non-scientific matters before it is time for the second round of drinks.

What topics should one cover during a meeting? Generally speaking, it is a good idea to avoid excessively long meetings; most are typically no more than an hour long. During that time, discussion can focus on many different things, such as the latest administrative restructure and how it will affect your lab's scientists, to discussions of the newest research just published online. I often ask a student or a fellow to present the progress of her latest research, usually accompanied by preliminary slides or figures of results that we can discuss together as a group. The development of new methods, or even how someone figured out how to handle a new analytical problem can all be useful and entertaining topics for a lab meeting. If a conference to which one or many participants will be attending is looming, lab meetings are great venues for practising oral presentations (Chapter 19). Post-conference debriefs are also a popular topic for lab meetings. I also like to include announcements of any successes, such as the acceptance of someone's article in a peer-reviewed journal, awards won, or successful grant announcements. Clearly, use your lab meetings to introduce new students or staff to the rest of the group, and give everyone a chance to meet and greet. The important thing here is to mix it up a bit, and not follow a set pattern with attendance, minutes, or forced contributions from all present.

On the matter of more social events that do not necessarily have much to do with the 'science' part of a lab, do not neglect these either. Whether they be formal holidays, celebrations of successes, birthdays, or just a regular pint or two at the local pub with some or all of the lab at the end of the formal working week, these can be integral for developing a healthy camaraderie amongst lab mates. In fact, some of my fondest memories of my previous labs are of these times. The take-home message is that you should not allow your lab members to fade into a solitary existence devoid of interaction with the rest of the membership. The more you foster social pollination, the higher your academic yields will be.

But a lab is much more than a series of regular meetings and social events. Most of your time will be spent working alongside your lab members in adjacent offices, and the larger your lab, the more organised you will need to become to schedule all of the professional interactions required (Chapter 16). Regular meetings with students, technicians, and fellows will be required on a case-by-case basis, with some people needing more time with you than others. A good strategy to maximise your own efficiency is to dedicate one to two days a week to such meetings, so that you minimise disturbances to your own writing and research time during the remaining days. I also endeavour to share the work load in this regard by formally including postdoctoral fellows on the advisory committees of students. This serves three functions: (*i*) it reduces the contact time between the primary supervisor (you) and the student, (*ii*) it often gives the student more time with an expert for learning laboratory, analytical, and writing techniques than they would otherwise receive from a busy lab head, and (*iii*) it provides early-career postdoctoral fellows with supervisory experience that they will need eventually to build their own labs. Win. Win. Win.

Of course, even the best, most functional lab will never be just smiles, camaraderie, and fun. No matter how prepared or proactive you might be, there will always be conflicts between some people in your lab. Most of the time this might be no more than a mild annoyance that has little effect on the lab's total social cohesion, but it can easily spiral out of control and turn into a battle that threatens the very fabric of the lab's existence. If a conflict comes to your attention, the best thing to do is to address it forthright by talking privately with each of the parties involved. If some sort of intervention is required, it is essential to identify it early and coerce the adversaries to achieve some sort of truce. The worst-case scenario is that you ignore rumours of conflict until they become intractable to resolve.

Conflict will not necessarily be restricted to other members of your lab; you could also find yourself as the target of ill will by a student or fellow, or you could take a dislike to someone who you believe is not pulling his weight. These things are inevitable, of course, but it is important to take note of any negative interactions such that if it comes to official, punitive action in the future, you are prepared to defend yourself. This might sound rather ominous, but it can happen to anyone, of any gender, and at any time. I have experienced some of these myself, and my colleagues collectively could write an entire tome about such negative experiences in the lab. These might include students who crack under immense academic or personal pressure, postdoctoral fellows who get involved in illegal activity, or technicians who believe they have far more intellectual property rights than might be justified. If you find yourself in a situation requiring outside intervention from say, a departmental head or your university's postgraduate office, then having a formal record of events will assist greatly in establishing an unvarnished chronicle of events. Take notes during strained meetings, and if you can, ask permission to record meetings for posterity (but do not do this clandestinely without the express permission of the people involved). And always take the moral higher ground by sticking to the business of science and avoid becoming mired in personality conflicts per se.

Personal involvement with lab members is another grey area that can lead to conflict, and in the worst case, even your dishonourable dismissal from your institution. As a first, overarching strategy, try to avoid forming close friendships with your lab members. While 'friendship' is itself a remarkably fluid concept that takes on many forms, I am referring specifically to close, personal friendships that border on the familial, rather than the day-to-day friendliness of casual interactions. At the same time, I admit that close friendships are also inevitable because of the high probability of meeting people of like mind during your professional life. In essence, I am merely offering a word of caution about soliciting friendships with your lab mates at the drop of a hat. The main reason I caution you is that if you eventually find yourself in a situation where some sort of punitive action is required, managing a lab charge-cum-close friend can be extremely uncomfortable. For example, let us say that a close friend who is also a doctoral student under your supervision is continually failing to achieve even modest milestones, something that would normally require hefty intervention on the part of the supervisory team. Critiquing a close friend in this manner is obviously awkward, and can even end up in the loss of friendship, the project, or both.

Friendship taken to the next level is even worse, and I am of course referring to intimate relationships between you and other members of your lab. It might seem unnecessary to address specifically here because it should be blindingly obvious, but do not, under any circumstances, copulate with your lab mates. As a lab head, your responsibility is to lead your charges onto becoming productive, independent scientists themselves, and mixing this business with your pleasure will nearly always end in (mostly your) tears. In fact, most institutions of higher learning have specific policies prohibiting such relationships, so you would do well to avoid acting on temptation in the first place. I know of too many circumstances where colleagues (both men and women) in a position of power have had relationships with students or staff, which then turned sour. The aftermath went well beyond mere broken hearts – they typically involved legal ramifications including everything from being barred from ever supervising students again, salary cuts, dismissal, court hearings, and even to imprisonment. Do not walk down this path. Even the perception of intimacy can be problematic even if nothing untoward eventuates; for this reason, make certain that your meetings are always done with the office door open. You do not want to give rumour a boost by hiding away somewhere with someone who either fancies you, or whom you fancy.

Intimate relationships amongst members of your lab is another thing entirely, and you should even expect them to occur. I have even had fellows who first met in my lab eventually marrying and having children together. This is of course none of your business, so be extremely careful about intervening here. The only time I would ever recommend intervention is when it becomes clear that the relationship threatens the scientific integrity, quality, or productivity of the people involved. This normally does not occur, unless the relationship ends while both people involved are still having tenure in your lab. Sharing an office with an 'ex' could end up being disastrous for the harmony of the lab in general, so be prepared to take control of the situation by separating those involved and nipping any aggressive interactions in the bud as soon as they happen. If things become too nasty, then you might even have to consider letting people go; but be warned, dismissing staff and students is a tricky, extraordinary course of action that favours the dismissed much more than those doing the dismissing. You will have to have a well-documented and water-tight case to be able to sack anyone, and of course, you will have to be beyond reproach yourself.

While I would not suggest that this next course of action is necessary for all, it is something to put on the radar for those who find themselves managing many scientists. As I indicated in the beginning of this chapter, scientists in general do not tend to be highly organised, and as you have most likely gathered by now, managing many people takes a good degree of organisational capacity. At the risk of giving scientists yet another task on the road to effectiveness, it might be a good idea for some of you to consider taking a management training course from time to time. There are probably many offered at your own institution, or via a professional organisation to which you might belong. Keep on the lookout for these, and discuss the need with your own administrative superiors. Tactics and styles that go into far more detail than I do here could further streamline your daily managerial routine such that your lab ends up operating like a well-oiled machine.

14 Making New Scientists

I am not exaggerating when I claim that a scientist is only as good as her postgraduate students and postdoctoral fellows, for this army of apprentices is what ultimately makes a laboratory – and the person who heads it – a success or failure. This is not to insinuate that the moral and professional duties of the academic scientist to mentor the next generation of scientists are any less important. But if more scientists realised that their success is utterly dependent on the quality of the people within their care, then more of us would have paid attention to good people-management skills earlier in our careers.

Students, and perhaps to a lesser degree, research fellows, are prone to figure out at some point during their term that they represent not only the wheels of the successful laboratory, but the axels, chassis, and engine as well. To extend this vehicle analogy, established and effective scientists spend most of their time merely steering the automobile in a particular direction and fuelling the vehicle with research funds. A poorly managed laboratory will therefore tend to make the students and fellows feel more like cheap labour or even slaves than as essential cogs, so it is imperative to give them responsibilities and opportunities beyond the typical tasks of the technician. A laboratory leader who writes all the grant applications, drafts all the manuscripts, decides on all the PhD projects, provides all the supervision, and in all other ways controls the academic lives of her protégés will probably not produce the next generation of effective scientists either. To be sure, it is a balancing act to promote independence and manage effectively one's apprentices simultaneously, so I can offer a few pointers about how you can achieve this.

THE DOCTORAL DANCE

If it has been a long time since you obtained your PhD, then it is perhaps easy to forget how vulnerable you were when you began navigating the path to academia. No two starting doctoral candidates are alike: some require excessive supervision, management, and scrutiny, while others prefer to do everything on their own terms. Neither type is ideal because the former requires too much effort to manage, and the latter misses essential supervisory guidance, so setting down a few rules from the outset is generally a good idea. Upon agreeing to supervise a new PhD student, I usually find it prudent to sit down and go over all of my expectations, as well as attempt to gauge those of the student. Establishing expectations from both sides of the relationship as early as possible can serve to avoid misunderstandings and conflict later.

While there are many different models of successful supervision, my own is to make it abundantly clear that I expect the student to publish (in peer-reviewed journals) all of the work arising from her thesis, and that circumstances beyond her control notwithstanding, ultimately all the results must be made public through publication. Each institution has its own official guidelines regarding intellectual property and the rights and responsibilities of the university, supervisor, and student, so I am not speaking of the legalities of intellectual property per se; rather, the initial conversation should establish the *culture* you want to perpetuate among all lab members. Be frank, honest, and supportive, and you will most likely have a mutually respectful and prosperous relationship.

A good way to avoid excessive interactions with students requiring abundant nurturing, or the prolonged absences of the more independent types, is to implement a regular reporting mechanism between students and their advisory committees. I have a simple approach that manages to work well in most situations, regardless of the student's peculiarities. This reporting consists primarily of a simple, one-paragraph, or bullet-point list of achievements made during

the interval between reporting times, as well as the immediate plans for the following interval. The interval itself can be set according to the needs and capabilities of both parties – I often suggest that the student sends a fortnightly e-mail with the required information to all members of his supervisory committee, or at least especially to me as the principal supervisor. This regular reporting serves many purposes, including: (i) the most obvious advantage of making sure the advisory committee is kept up-to-date with the student's progress, (ii) encouraging the student to keep organised by specifying accomplishments as well as setting near-term goals, and (iii) maintaining an interaction, albeit virtually, between the supervisors and student when face-to-face meetings are difficult or impossible. I would never suggest that this action replaces real, face-to-face meetings; instead, the reports can maintain a dialogue in the interim, as well as provide warning signs of inactivity, unproductive lines of inquiry, or other problems that could hamper a student's progress to completion. I have been able to rescue more than one stray analysis or prevent weeks of wasted time by reviewing these e-mails from my students.

A good supervisor should also be prepared for a PhD student's meltdown, which I estimate occurs typically around the 50–60 per cent completion mark of the thesis. Almost without fail, my students will hesitantly request a meeting with me around this time, during which they will divulge a deep feeling of incompetence, a profound doubt about their chosen career path, and a terror of not being able to complete in time. The circumstances are always slightly different, but the main message is the same. I have come to realise that this is the rule rather than the exception, and that it can be viewed in some ways as a necessary rite of passage. My strategy to deal with this form of doubt is to let the student download his worries without fear of being interrupted or belittled by me. In fact, during this part of the conversation I try not to say much at all, instead giving the student the liberty to speak at will. Then, without any hint of judgement, I attempt to point out that (*i*) all students go through this, (*ii*) it is perfectly acceptable to feel like this (I did too), (*iii*) a PhD is difficult, otherwise the diploma would not be worth the paper on which it is printed, and (*iv*) you (the student) know more about your topic than anyone else in the world. This last point might seem overly generous, but I maintain that a PhD student who has progressed to complete over half of her dissertation is necessarily a world expert in her particular (admittedly narrow) field of research. No one but a PhD student is consumed so completely by his project, meaning that he has more of that specific information at his fingertips than anyone else. This useful realisation alone can be enough to lift the student out of her mental funk, and all but guarantee a successful completion.

EARLY TO PRESS IS BEST FOR SUCCESS

On the subject of a publication 'culture' that I foster within my lab, I want to reiterate that this should be tailored to suit the particular philosophy of the lab head in question. If you already lament the emphasis on publication for publication's sake, then you can take this information with a grain of salt. While I would tend to agree generally with a person holding this opinion, I also argue that the realities of modern science make this emphasis inescapable; I am not aware of a single scientist known for her important scientific contributions who does not have a prolific publication output. No, publishing bucketloads of papers will not guarantee you a Nobel Prize, but if you do not publish, you will most certainly not win one either. Accepting therefore that effective scientists must publish, and do it often and (mainly) in quality journals (see Chapter 6), then there is empirical evidence to support the notion that instilling this culture early on is essential.

Publication frequency is therefore correlated with 'success' (however you might wish to define this; see Chapter 1), even if it is not the perfect indicator. So, what factors lead someone to publish more than someone else? There are a few possibilities here, with some well-known mechanisms, and others that are only suspected. Using the *curriculum vitae* of 1400 biologists in various disciplines (excluding medical research) from four different continents, my colleagues and I measured the number of peer-reviewed publications that each scientist in the sample had written by the time each had completed her or his PhD, as well as the number of publications she or he had published ten years later (3). We also collected information on the scientists' gender, whether English was their first language, and the international ranking of the university where each obtained the PhD.

Combining the data into a series of linear models, we asked the following questions:

(i) Given that our sample included people who stayed in science for at least ten years (i.e., we did not include people that gave up their scientific careers in the interim), do male scientists publish more than female scientists?

(ii) If you went to a highly ranked university to obtain your PhD (e.g., Cambridge, MIT, Oxford, Harvard, University College London,

Copenhagen, Stanford, Melbourne, etc.),[1] are you more likely to publish at a higher frequency than someone who received her PhD from a lower-ranked institution?

(*iii*) Most scientific results are published in English these days, so if English is your first language, do you have an advantage, and therefore publish more than someone for whom English is a second (or third, fourth, ...) language?

(*iv*) If you start publishing early in your career, does that set the pace for the rest of it?

Unless you are already familiar with this research (3), I am sure that most of you will be pleased to learn that the most important determinant of one's 'long'-term (ten-year) publication success (frequency) is how many papers a scientist has written by the time she has completed her PhD (Question Number *iv*). This effect increases markedly if we take the number of papers published three years after PhD completion as a predictor. To make the point again that publication output is a reasonable metric of 'success', we also found that it was highly correlated with the person's ten-year *h*-index, which is an index of one's citation history (see also Chapter 6).

As for the other questions, we did find some minor effects. Even after removing the well-known 'attrition' effect of female scientists (i.e., women leaving their science careers earlier than men), men tended to publish a little more than women (Question Number *i*). There are many potential reasons for this, including still largely male-dominated academic and publishing systems, misogyny, and the additional constraints of child rearing (see more in Chapter 15). However, while having English as a first language gave scientists a slight publication advantage (Question Number *iii*), the effect was

---

[1] There are several systems that use different criteria for ranking universities, including the *Academic Ranking of World Universities* (shangairanking.com), the Times Higher Education *World University Rankings* (www.timeshighereducation.com/world-university-rankings), the Webometrics *Ranking Web of Universities* (www.webometrics.info), and the Quacquarelli Symonds (QS) *World University Rankings* (topuniversities.com).

extremely weak. Possibly one of the most interesting results was that PhD-university ranking (Question Number *ii*) had absolutely no discernible effect on publication output, regardless of which ranking metric one uses.

There are a few important messages arising from this analysis. First, if you are a PhD student or an early-career researcher, it is important to put the effort into getting those first papers out. In other words, early to press is best for success. Second, if you are considering people to hire for a new position and you are taking a gamble on their potential to publish, you should perhaps place a stronger importance on their publication output to date (all other considerations being equal). However, employers should *not* choose men over women, nor should they blindly hire people with English as a first language, merely as a means to maximise potential publication output. The differences were so minor that they would be overwhelmed by other considerations of the applicant, so even if one was so morally bereft as to favour native English-speaking men, it would do no good. Further, not only were the gender and language effects weak, they nearly disappeared once we considered the data three years after PhD completion. Of course, our results were derived from a sample of (non-medical) biologists, so they might not necessarily apply, at least in the same ways, to other disciplines of science. Still, even in the absence of hard evidence I would be surprised if similar patterns were not evident in other science disciplines.[2]

My final recommendation here is that if you find yourself in a position to hire early-career researchers, perhaps considering applicants who have recently completed their PhD for a postdoctoral position, you should not favour those who have a 'better' pedigree in terms of university rank. Indeed, a person who obtained her PhD from a lower-ranked university, but has more publications than another person who went to a highly ranked university, has a higher likelihood

[2] The social sciences could be another matter entirely.

of delivering the academic goods later on (all other things considered being equal, of course). Perhaps students (and their parents) should also put less emphasis on university ranking and more on the people with whom they will be working when considering where to do their postgraduate studies.

15 Human Diversity

I have discussed quite a few aspects of *how* an effective science lab can be run, but so far, I have neglected to define *what* a lab is. My definition of a 'lab' (or 'centre', 'group', 'research unit', 'institute', etc.) in this context is simply a group of people who do the science in question – and people are a varied bunch, indeed. I wager that most scientists would not necessarily give much dedicated thought to the diversity of the people in their lab, and instead probably focus more on obtaining the most qualified and cleverest people for the jobs that need doing. For example, I have yet to meet an overtly racist, sexist, or homophobic scientist involved actively in research today (although unfortunately, I am sure some do still exist), so I doubt that lab heads consciously avoid certain types of people when hiring or taking on new students as they once did (74, 75). The problem here is not that scientists tend to exclude certain types of people deliberately based on negative stereotypes; rather, it concerns more the *subconscious* biases that might lurk within, and about which unfortunately most of us are blissfully unaware. But an effective scientist must be aware of, and seek to address, his hidden biases.

It is time to place my cards on the table: unless you have not yet noticed, I must reveal that I am a middle-aged, Caucasian, male scientist who has lived in socially inclusive and economically fortunate countries his entire life. As such, I am the quintessential golden child of scientific opportunity, and I am therefore also one of the biggest impediments to human diversity in science. I cannot apologise for being these material things, for am not able to change my status[1] per

---

[1] Technically, I could alter my gender, but I am auspiciously content to remain in the morphological form in which I was born.

se; however, I can change how I perceive, acknowledge, and act to address my biases. The earlier a scientist recognises these challenges in her career, the more effective she will be. I will deal with some major categories of potential bias, and outline how one might become at least a little more aware that they exist to avoid perpetuating the *status quo*.

GENDER BALANCE

I acknowledge that as a man, I am stepping onto thin ice even to dare to discuss the thorny issue of gender inequality in science today, for it is a massive topic that many, far more qualified people are tackling. But being of the male flavour means that I have to, like an alcoholic, admit that I have a problem, and then take steps to resolve that problem. After all, privilege is generally invisible to those who have it. If you are a male scientist reading this now, then my discussion is most pertinent to you. If you are female, then perhaps you can use some of these pointers to educate your male colleagues and students.

There is now ample evidence that science as a discipline is just as biased against women as most other sectors of professional employment, even though things have improved since the bad old days of scientific old-boys' clubs (74, 75). We now know, for example, that journals tend to appoint more men than women on their editorial boards, and that editors display what is known as *homophily* when selecting reviewers for manuscripts: the tendency to select reviewers of the same gender as themselves (76). Likewise, experimental evidence demonstrates that scientists in general rate male-authored science writing higher than female-authored works (77), and that academic scientists tend to favour male applicants over females for student positions (78). In the United Kingdom, as I suspect is more or less the case almost everywhere else, female academics in science, engineering, and mathematics also tend to have more administrative duties, and hence, less time to do research; they also have fewer opportunities for career development and training, as well as earning

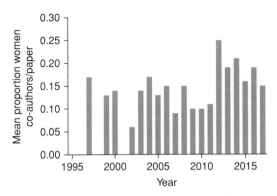

FIGURE 15.1 Mean proportion of my female co-authors from 1995 to 2017.

a lower salary, holding fewer senior roles, and being less likely to be granted permanent positions (79).

It might be tempting to tell yourself that, of course, these general trends are not because of your own biases, just of the scientific community in general. This is called denial. While the first step is to accept that gender bias does exist in science, you should at least entertain the notion that you are biased as well. My own epiphany in this matter came in the form of a self-evaluation of my publishing trends after a colleague suggested that I should perhaps examine the ratio of my male to female co-authors. My preconception was that while I was cognisant of the dominance of men as my co-authors, I was convinced that I had a 'healthy' gender balance. I was wrong (Figure 15.1).

A casual glance could suggest a weak increase in the proportion over time (a positive sign, to be sure), but the extent of the y-axis says it all – most (between 75 and 90 per cent) of my co-authors are men. Whether this range matches the gender ratios amongst publishing scientists that I could use for comparison in my field is unknown, simply because such data have not been widely published. Regardless, such data do not testify to achieving gender equality any time soon.

These results therefore challenge my innate presumption that I have not subconsciously tended to choose male colleagues (including students and postdoctoral fellows) over female ones. In other words,

I am guilty of passive sexism (or technically, *homophily*). And such bias is not restricted to the old, white men of science. Recent evidence from one university clearly shows that male undergraduates underestimate the academic performance of their female peers in biology classrooms (80), and any quick glance at the gender ratios of senior management of nearly all universities around the world will confirm that the men are overwhelmingly dominant numerically.

A cold, hard gaze in the proverbial mirror is therefore not only recommended, I argue that all scientists must do it, and not merely because it might 'look good' to your community of peers. In fact, there is now a growing body of evidence to demonstrate that gender equality is good for all involved, *including men*. Overall, a better gender-balanced workplace leads to higher employee happiness, satisfaction, and even health (81), and men are less prone to violent behaviour when gender equality is promoted (82). Given that it is in the lab's best interest to be composed of a group of healthy, satisfied, happy, and non-violent scientists, moving towards gender equality is only ever going to improve your lab's efficiency, productivity, and effectiveness. Furthermore, there is evidence from the corporate world that better gender balance stimulates higher innovation, and innovation is certainly a desired quality of a productive science lab (83). Giving your students and fellows a more gender-balanced workplace will therefore provide an environment more conducive to scientific output and minimise the likelihood of social problems arising within the day-to-day dynamic.

Of course, there is no empirical evidence whatsoever to support the rather outdated and sexist idea that the genders differ in their capacity to do science and mathematics, despite some lingering opinions to the contrary. I will add to this that even if certain components of the cognitive *smörgåsbord* required to be an effective, broad-thinking, creative, and innovative scientist are ever shown conclusively to differ in any respect between the sexes, then it would only make sense to a forward-thinking lab head to promote gender diversity. As I argue throughout this chapter, diversity is an essential ingredient to bring out the full potential of your lab, because different experiences, capacities, mindsets, and points of view all add their uniqueness to the innovation soup of your science direction and output.

Understanding that gender inequality exists, and appreciating that it can harm your lab's productivity, do not necessarily translate to fixing the problem automatically, so what can one do to improve the *status quo*? The solution requires more than just trying to be fairer in the process of hiring and student appointments; in fact, a lab head can do so much more. The first cab in this rank is simply encouraging your lab members to have regular discussions of gender issues, including their effects, how to address inequality, and perhaps ideas for promoting equity. At the same time, it is imperative that you identify people who actively downplay the importance of gender inequality in science and show them how improvement benefits everyone, men included. If active sexism does rear its ugly head, it is equally important to avoid turning a blind eye and instead speak out against the perpetrators. That goes too for pointing out incidences of stereotypes that could be casually dropped without the awareness of the culprits. Going through self-assessment exercises[2] designed to quantify any subtle and subconscious biases that might exist is also a good idea.

Some have argued that the principal demons of sexism have largely disappeared from science and that social constraints are now more important for maintaining gender inequality (75). While I am not convinced this is necessarily always the case, the point that for many women the very real additional effort required to have and raise children in a happy family setting is still today a major determinant of many women choosing not to enter, electing to leave, or not having sufficient time to engage in a science career (75, 84, 85). As such, anything one can do to make a woman scientist's life easier in this

---

[2] One way you can do this is to take the Harvard Implicit Association test for Gender (implicit.harvard.edu/implicit).

regard is a good idea, from simply encouraging children to being in the lab when and where possible and appropriate, to making child-care options more accessible.[3] Family-friendly workplace environments do, as the evidence above suggests, enable scientists in general to do a better job, and allow women scientists in particular to hold their positions for longer and advance to higher positions in the academic hierarchy more rapidly (86). Even outside of the lab, such as when organising and attending conferences, there are many ways to make things more family-friendly for women scientists, such as having specific gender-equality policies, ensuring balanced conference committees, establishing gender-related (and other) codes of conduct, providing child care, and supporting women scientists financially (87, 88).

CULTURAL DIVERSITY

Many of the same problems underlying gender inequality, as well as actions to lessen its incidence, also apply to cultural inequality; however, cultural inequality is a more complex issue, and not merely because there are rather a lot more different types of culture than gender. With all the nasty nationalism and xenophobia gurgling nauseatingly to the surface of our political discourse these days, it is worth some reflection regarding the role of multiculturalism in science. I am therefore going to have a crack at addressing this here.

Just like few scientists are overtly sexist, not many would elect to hold up their hand and claim that they are racist. In fact, most scientists are of a liberal persuasion generally, and tend to pride themselves on their left-wing political tendencies. In other words, we tend to think of ourselves as dispassionate pluralists who only judge the empirical capabilities of our colleagues, with their races, genders, sexual persuasions, and other physical attributes irrelevant to our

[3] In many countries, including my own, child care is a state-subsidised service that is often available at the institution itself. Regardless, there is usually a cost, so a lab head can potentially negotiate salary packages to minimise the financial burden if necessary.

assessment. We generally love to travel and interact with our peers from all nations and walks of life, and we regularly decorate our offices and homes with cultural paraphernalia different to our own. But are we as unbiased and dispassionate as we think we are? Do we take that professed pluralism and cultural promiscuity with us to the lab each day? As the previous section regarding gender confirmed, perhaps we could, and should, do better in the multicultural arena as well.

I know from personal experience and from casual discussions with colleagues that it is not always the path of least resistance to take on a student or research fellow from overseas. For one, the educational and funding system – for all its professed emphasis on internationalism – is stacked against the academic scientist. Finding a scholarship that will pay both the living expenses and foreign-student fees of a prospective foreign student is decidedly more challenging than obtaining one for a citizen in most countries. Such scholarship opportunities do exist, and if the student is good enough, there is nearly always a solution, but the entire process represents a good deal more effort to make it happen compared to finding a student from the neighbourhood.

Then there is the issue of language capability. Many scientists complain that it is not their role to tutor a student or a research fellow in the subtleties of the English language, and that poor writing skills in English hinder their capacity to produce good scientific publications. I agree that it can be an additional challenge (see Chapters 2–7), but I disagree entirely that having English as a first language these days provides much in the way of an advantage for scientists in training. As I discussed earlier in the book, the writing capacity of even native English speakers is on the wane, so in my experience there has been no more effort to get a non-native speaker's English prose up to scratch relative to the native English speaker's. In fact, some of my best writers have hailed from countries where English is not the native tongue. In addition, as our previously discussed (Chapter 6)

analysis of publishing trends showed, neither first language nor gender explained much variation at all in the publication output of young scientists[4] (3).

The perceived and possibly real disadvantages of taking on foreign charges are therefore weak at best, which begs the question whether the desire to increase the cultural assortment of your lab is a good thing to do for your science effectiveness. The short answer is *yes*, for as I discussed in the previous section, a lab with a varied mix of experiences, knowledge bases, perspectives, genders, insights, and values is necessarily going to catalyse the ideas bubbling away in your lab's crucible of innovation. Uniformity breeds staleness, inhibits transdisciplinarity,[5] and quells novelty. If you want to be at the cutting edge of your scientific endeavour, then a diverse, multicultural, and linguistically variable lab will help you get there. Let the right-wing, populist xenophobes vomit their racist bile all that they want, while you quietly get on with the job of making the world a smarter, more innovative, multicultural, understanding, and collaborative place.

DEALING WITH REQUESTS FOR SUPERVISION

Accepting the idea that multiculturalism is a laudable objective for your lab's composition, the next challenge is finding ways to achieve it. The irony here is that most university-based academics regularly receive requests from people around the world wishing to be considered as prospective postgraduate students, mostly to do a PhD. I probably receive an average of three to five such requests per week via e-mail, as do many of my colleagues. Unfortunately for those making the inquiry, I trash most of them almost immediately from my inbox folder. My behaviour could be interpreted as perhaps a little unkind, but if you were privy to reading these requests it would become

---

[4] Biologists, in this case.

[5] A term I use to describe *the cross-pollination, uptake, and evolution of techniques and ideas from different science disciplines*, which differs from *multidisciplinarity* – people with different expertise working together to solve a complex problem.

painfully clear that few of these people had given much thought to their inquiry, or how they might be perceived by the intended recipient (in this case, me). To give you a flavour of what I mean, most of the messages go along the following lines:

> Dear Proffesor Mr Corey Bardshaw,
>
> I wish to write you to seek for supervising towards PhD degree. If you not intersted, assist me to get other supervisor.
>
> Sincerely,
> Astudent W.H.O. Isnotfromaroundhere

I am sure you will agree that the curtness, lack of detail, bad grammar, and the minefield of spelling mistakes[6] would not flush you with a sense of confidence about the student's potential capabilities. You might therefore appreciate more why these e-mails end up in the trash so quickly. I am not, however, naïve enough to think that most of these are serious requests for supervision; indeed, many of them seem to be desperate cries for help to assist someone with quitting his country of origin, for reasons that have nothing to do with academic pursuits. In other words, these people appear to be using academia as a 'refugee' tool[7] rather than trying to find the best supervisor to suit their academic interests. It can be tempting to engage with correspondents who provide a little more detail, but for the sake of winnowing the refugees from the genuine student prospects, and for any readers who are themselves contemplating contacting potential supervisors overseas, I provide the following components of an ideal first-contact e-mail message.

---

[6] Including my own name; I mean, really – how difficult is it to get the supervisor's name right? In fact, one recent requestor addressed me by the formal name of my Chair (it was 'Sir Hubert Wilkins'). Sometimes, you really must wonder how some people have enough common sense even to turn on the computer.

[7] The inverted commas imply that I do not necessarily mean 'refugee' in the sense of someone who is persecuted for political or religious reasons and forcibly displaced from her state of origin. Instead, I am applying the term rather liberally to mean someone who, for whatever reasons, would rather not live where they were born and brought up.

Perhaps most importantly, it is essential that the seeker has clearly identified someone with approximately the correct expertise in the area of interest. I have received hundreds of e-mails from prospective students requesting supervision for projects in electrical engineering, agronomy, cellular function, or toxicity trials, yet I do not research or have any expertise in these fields whatsoever. While I like to claim that I am something of a science generalist, these people are stretching the fabric of reality a little too far. Second, if the e-mail looks like a form letter, it probably is a form letter. In other words, if I cannot tell that the person has specifically targeted me for my specific expertise, reputation, and track record, I am unlikely to give that person a second glance. If the person has not taken the time to peruse the prospective supervisor's online curriculum vitae and invest even a few sentences on how her interests align with your own, it will be clear. A lack of researching capability at this stage does not bode well for the future endeavours of a research student.

The seeker should also have a clear idea of the area she wants to research for the PhD. Writing nothing more than 'I am interested in physics' to a particle physicist working at the CERN[8] is woefully insufficient. The student must (again) do some actual *research* into a question that is not only interesting to the potential supervisor, it must also represent a question that is worth investigating in the field. In other words, it is essential that the seeker has done some background reading and identified the 'next big things' in the specific field. Of course, I would never expect a complete dissertation proposal at the initial stages of contact, but I do expect the prospective student to know roughly what hypotheses she would like to test.

In terms of format, nothing says 'incompetent' more than a single-line e-mail, and a lack of formality or politeness is not a great way to get off on the right foot. Similarly, while I do not expect my students to have perfect English when they start with me, especially

---

[8] The acronym was originally derived from the *Conseil européen pour la recherche nucléaire*, but which is now (in English) the 'European Organization for Nuclear Research' (cern.home).

those for whom English is a foreign language, I *do* expect them to have put in the effort to avoid gross grammatical or spelling mistakes in the initial contact message. If they have not bothered to check their English with a colleague or friend, then it possibly demonstrates laziness or a lack of attention to detail, which are not the best character traits to have if one is trying to become a scientist. Do not dismiss the contact entirely if there are only one or two minor spelling mistakes, for this can happen even to the best writers of English.

Many of the e-mails I receive also include an unsolicited curriculum vitae as an attachment, which might encourage the inexperienced prospective supervisor to investigate further. However, I do not usually bother opening the attachment until I have first had a more formal and positive interaction with the seeker. Ideally, the prospective student should initiate contact and, following the recommendations above, enter into a dialogue with the person he is targeting. Once the online relationship and dialogue have been established, only then should the potential supervisor request a more detailed curriculum vitae. Finding a PhD supervisor is not like applying for a job – it is a personal relationship with an academic mentor and ideally, a mutually respectful one at that.

Another good sign is if the student has provided at least one example of her previous scientific writing experience. The ideal here is a previously published scientific article, whether it be first-authored or not, but even a technical report, a lay article, or something similar can testify at least partially to the student's academic potential, and to your futurity sanity when supervising them. Finally, I tend to look for some evidence that the overseas student has at least given some passing thought to how she might finance the whole endeavour. Of course, it is typically not the student's primary responsibility to find his own scholarship and living allowance, but different academics, institutions, and countries have variable capacity to guarantee such financial support for incoming students. Scholarships are not unlimited, but there are usually enough of them kicking around that the best students can obtain. But if the seeker e-mailing you merely asks that you fund him outright, without any indication that he has at least an inkling that you are not an untapped bank account, then my opinion is to raise an eyebrow about that potential student's ingenuity. However, if she has done some preliminary cyber-searching for scholarship opportunities available to her without your prompting or encouragement, the promise of capability is higher. The availability of and eventual success in securing funding is often the longest delay in appointing a student to postgraduate research, so the seeker should also demonstrate some idea of how long the process might take. Demonstrable impatience is not a good look.

I certainly do not want to give the impression that we academics are a narcissistic lot with nothing better to do than demand supplication from unworthy sycophants. Nothing could be farther from the truth. We absolutely depend on good students, as much as good students depend on us to guide them to academic independence. As an active academic scientist, I am constantly on the lookout for high-performing, intelligent, and hard-working students, so I relish being contacted by interested prospects. If the seeker turns out to be as good as she claims, then I will gladly do all that I am capable of doing to find her the appropriate position, topic, and funding; however, more often than not, the initial contact message merely demonstrates that I should not even entertain the idea. In other words, be selective.

SEXUAL DIVERSITY

I have claimed throughout the book that we scientists are more or less the same as everyone else, albeit perhaps with a slightly higher tendency towards minor personality disorders (Chapter 13) and a sometimes socially awkward manifestation of hyper-curiosity and scepticism. Those differences aside, we are as human as the rest of the population, and so we also cover the gamut of personality types, interests, and sexual orientations. If anything, academia is probably a slightly safer haven than most other professional arenas for those scientists who fall outside the central tendency of the sexual-interest axis. If you occupy the approximate middle of that particular distribution, then do not be surprised that many of your students, postdoctoral fellows, and colleagues do not. While I am not certain of the sexual orientation of most of my colleagues (nor do I particularly care because it is really none of my business, nor is it something that would alter my opinion of them), I personally know two transgender scientists in my field, and LBGTQ[9] students, research fellows, and collaborators are ubiquitous in my general research area.

If women scientists are still finding that acceptance and equal opportunity remain elusive, imagine what many LBGTQ scientists go through every day. While conditions are undoubtedly improving, even the most progressive societies are still a long way off from being completely open, accepting, and tolerant of non-standard sexual orientations and behaviours, just like we have a long way to travel to address even today's remaining gender inequalities. I cannot profess to offer any solutions to these inequalities and intolerances, except to encourage you reading this to examine your own prejudices regularly and earnestly. As a supervisor, it is a good idea to be aware that some of your students might be finding that the workplace environment is less accepting than it should be, and if so, you should identify to them that you are 'in their corner'. If there are clear signs of conflict amongst people in your lab, department, or faculty related to sexual orientation, then it is also your responsibility to nip them in the bud as soon as possible. In the university setting anyway, there are usually organisations or services available to help resolve such conflicts and to provide support for people who feel victimised, and a responsible

---

[9] I apologise in advance if I have not used the most conventional initialism here, for I cannot claim to be an expert in the appropriate identification terminology. For those who are still not familiar with this initialism, it stands for: *lesbian*, *bisexual*, *gay*, *transgender*, *queer* (or, *questioning*). Although I am told by some that even this term is falling out of fashion as people increasingly shy away from stereotypical categories and their labels.

lab head will do her utmost to be aware of these facilities. Inappropriate sexual behaviour excepted (see Chapter 13), the take-home message is to be as open, accepting, and supportive of your colleagues, students, and co-workers as possible, no matter what or who they might prefer.

16 Splitting Your Time

Having read up to here, I suspect some of you are wondering how the hell you are going to balance all the requirements of an academic life in science. From actually doing the science (experimentation, field collection, lab work), analysing the data, writing papers about your results, reviewing, writing grants, to mentoring students – not to mention trying to have a modicum of a life outside of the lab (see Chapter 17) – it is forgivable to feel more than a little daunted. As I have previously mentioned too, many scientists, including me, are not innately well-organised, so the challenge is typically even more overwhelming than it initially appears. While there is no empirical formula that makes you run your academic life like a well-oiled machine, I can offer a few suggestions that might make your life just a little less chaotic.

The elephant in the room of finite academic time is of course the dichotomy of teaching versus research. I admit here that I have no additional insights to what many academics and university administrators have spent a long time contemplating, reformulating, trialling, and ultimately failing to resolve. In fact, the teaching–research nexus has been problematic ever since humanity had the bright idea to invent the first institutes of higher learning, so I fear it is a perennial issue. Academics have always struggled with finding a balance, and so it is safe to predict that they will continue to do so long after my bones have turned to dust.

Some of the riddle will be solved for you before you can even begin to establish your own teaching–research equilibrium, because most beginning academic contracts clearly stipulate the proportion of your 'full-time equivalent' that you must spend lecturing students. The proportion for newly appointed, just-of-the-postdoc-stage academics varies from institution to institution, and country to country, but teaching in some form typically ends up taking most of your time. In some countries, teaching itself is the bedrock upon which an academic career is built, so the traditional 'tenure track' means that the academic must establish a strong teaching portfolio such that they can ultimately be granted a state of permanence within their institution of higher learning; this ironically tends to mean a reduced teaching load. In other institutions, teaching does not guarantee any sort of permanence, nor is it valued as much as research. I suspect too that the traditional concept of tenure is on the wane globally because of the intense competition from the great pool of talent that our universities are pumping out each year.

Whatever your situation, an entropic process of randomised sorting ensues, with the more gifted teachers precipitating towards a dominant lecturing role (thus displacing some, or even all, of their research), whereas the less pedagogic emphasise their research role (but this time, displacing the teaching). The division is never comfortable, and administrators will inevitably pressure their charges towards parity between these two roles. Whether you are a better 'researcher' or a better 'teacher', get used to someone telling you to do more of the activity you like to do the least. However, as one climbs the slow and cumbersome academic ladder, the balance can become more harmonious, such that the more research-focussed can choose to teach less by 'buying out' their teaching obligations with grant monies for example, or the more teaching-focussed can justify doing less research by winning accolades highlighting their teaching prowess and influence. If you are one of the lucky few who manages to hit that magic balance between the two, then to you I earnestly tip my hat. If I can sum up any advice on this matter at all, is that you should steel yourself to the near-inescapable struggle.

Recalling earlier chapters in this book, I hope I have established the necessity of at least some high-quality research output to maintain an effective scientific career. In that respect, I lay my biases openly on the table. If research is not your desire or forte, then

perhaps this book is not necessarily for you. If you are like most of your colleagues and must split your time between these two main preoccupations, then there are some considerations to assist you. As I also stipulated earlier, your peer-reviewed journal articles will lead to citations, which will themselves bolster your reputation and assist you in obtaining funding so that you can do more research, and write more articles, to obtain yet more citations, and so on. I have been convinced that while this never-ending, circular existence might seem superficial and unfulfilling, it is a formula that will increase your capacity to be independent and *choose* which aspects of scientific academia you would rather spend most of your time doing. Research output is therefore your currency to academic freedom.

With this in mind, I have over the course of my career drawn up a type of priority list that I try to follow as closely as the vagaries of academic life will allow. I am not foolish or arrogant enough to believe that my list is any better than anyone else's, so the essential lesson here is that the effective scientist should endeavour to construct one that works for her. Think of the list not so much as a linear, step-by-step pathway to academic freedom; rather, view it more as a fluid concept to help guide you in your day-to-day activities. Your list will also mutate and self-adjust from time to time as the tides of opportunity ebb and flow (e.g., many funding agencies have specific deadlines, so academic scientists often refer to the time leading up to these as 'grant season').

PRIORITY 1: REVISE ARTICLES SUBMITTED TO HIGH-RANKED JOURNALS

Barring a family emergency, my top priority is always revising an article that has been sent back to me from a high-ranking journal for revisions. This is low-hanging fruit on the publication tree, for you have already scaled the most difficult branches to get to this stage. Bite the bullet and spend the necessary time (no matter how awful and nasty the reviewers' comments; Chapter 7) to complete the necessary revisions and return the updated manuscript to the journal in all haste.

PRIORITY 2: REVISE ARTICLES SUBMITTED TO LOWER-RANKED JOURNALS

In reality, I could have lumped this priority with the previous, but I think it is necessary to distinguish the two should you find yourself in the fortunate position of having to do more than one revision at a time. The only difference is that the previous will tend (but is not guaranteed) to attract more citations once published, so it should be placed at a slightly higher priority.

PRIORITY 3: EXPERIMENTATION AND FIELD WORK

Clearly, most of us need data before we can write papers, so this is high on my personal priority list. If field work is required, then obviously this will be your dominant preoccupation for sometimes extended periods. Many experiments can also be highly time-consuming, while others can be done in stages or run in the background while you complete other tasks.

PRIORITY 4: DATABASING

This one could be easily forgotten as you pull up the chair to your desk in the morning, but it is a task that can take up a disproportionate amount of your precious time if you do not deliberately fit it into your schedule. Well-organised, abundantly meta-tagged, intuitive, and backed-up databases are essential for effective scientific analysis (see Chapter 11), so second only to generating the data is the process of getting them into a useable and easily retrievable format. Databases are increasingly complex these days as new technologies unleash mind-bogglingly huge data streams at the mere flip of a switch. Even with dedicated and full-time database managers, an academic scientist now has to be across a wide range of databasing capabilities and an increasingly sophisticated array of networking technologies that will only intensify in the future. Good data are useless if you cannot find them or understand to what they refer.

PRIORITY 5: ANALYSIS

Clear scientific results do not just jump out of the computer at you, even if you are the best database expert around. Good analytics and I argue, coding skills (Chapter 10), are today essential tools in the effective scientist's toolbox. In terms of the time spent on each element of producing a scientific article, I rank data analysis as the third-most demanding stage of the process (following experimentation/data collection and databasing). Of course, analysis demands vary amongst, and even within, the scientific disciplines, but concomitant with rising data availability and complexity are the increasing demands of more and more complex analyses.

PRIORITY 6: WRITING ARTICLES

Despite my emphasis on the actual process of writing scientific articles, I have placed the *writing* phase in this position because one generally cannot write much without data or analysis.[1] And just like you should perhaps prioritise revising manuscripts from higher-ranked journals, so too is it a good idea to prioritise writing manuscripts you intend to submit to the higher-ranked journals in your discipline. With maximising citations as a soft target, it should be obvious how to do this if you have competing manuscripts awaiting completion.

PRIORITY 7: CONTRIBUTING TO AND EDITING YOUR COLLABORATORS' MANUSCRIPTS

If you recall in Chapter 4 my recommendation about designing your manuscript-writing protocol to ensure that your collaborators contribute in a timely fashion, your well-prepared collaborators will also likely set deadlines for you to provide comment on the manuscripts they are leading. Of course, these demands will necessarily shift around your priority list accordingly. However, in terms of personal choice, I nearly always prioritise commenting on or editing one of my student's manuscripts relative to all others, even if it concerns an article targeting a lower-ranked journal. I do this for four main reasons: (*i*) it teaches the student to value a rapid turn-around ethos, (*ii*) it minimises the time it takes a student to publish (with the clear advantages this approach provides; Chapter 6) (3), (*iii*) it nearly always ends up increasing one's own publication track record, and (*iv*) a good supervisor is morally and ethically bound to facilitate the student's transition to academic independence by prioritising feedback. Thus,

---

[1] Conceptual perspectives, non-systematic reviews, and comments notwithstanding.

I always spend time on a student's manuscript before doing the same with another colleague's.

PRIORITY 8: MEDIA ENGAGEMENT AND WRITING PRESS RELEASES

I can understand if you think this one is a little over-prioritised in my grand list, and you certainly have the right to disagree. I do argue however that effective engagement with the media is now more than it has ever been an essential component of becoming an effective scientist. While I will dwell on this subject more in Chapter 22, it suffices to emphasise two points about media engagement: (*i*) it generally takes little time relative to your other duties, and (*ii*) one should never underestimate the corollary benefits of effective public engagement with respect to your academic reputation as well. Writing a press release, especially with the help of the media professionals at your institution (Chapter 22), takes little time, and the potential benefits of doing so vastly outweigh the loss of perhaps a few days at most to do other, more pressing tasks.

PRIORITY 9: STUDENT AND POSTDOCTORAL FELLOW MEETINGS

My students, and my staff, are always high on my priority list for the reasons I outlined in *Priority 7*. Some scientists prefer to block out specific days of the week for such meetings to maximise their time efficiency, while others opt for a more *ad hoc* approach. Regardless of your preferred strategy, avoid falling into the enticing trap of putting off these meetings for too long. You can save both the student/fellow and yourself months of future headaches just by being proactive and nipping bad ideas or unproductive lines of scientific inquiry in the bud before they progress too far.

PRIORITY 10: WRITING GRANT PROPOSALS

As I mentioned above, grant-writing 'seasons' will shift this priority up or down accordingly. However, planning to write proposals well in advance of the deadline is universally a good idea, even though I know that most of you will not heed this advice. While I would not suggest that grant writing should take priority over preparing your scientific manuscripts, at some point you will have to buckle down and just write the proposal.

PRIORITY 11: PREPARING AND DELIVERING LECTURES

Again, this priority will shift according to your circumstances, but if you value the balanced teaching-research model, then this necessarily belongs lower down on the list than writing activities per se. Like writing grant proposals, preparing lectures can take a hefty amount of your time, especially if you are starting from scratch. I generally bank on ten hours of preparation for every hour of lecturing I do (in the case of lectures that I have never given before).

PRIORITY 12: ATTENDING SEMINARS AND CONFERENCES

When you are pressed for time, attending a departmental seminar might not be all that appealing. These tend to occur in the middle of the day, or just before you head home for the evening, so they often break the flow of your daily routine. That said, I wager that seminar attendance has dwindled over the course of my career merely because demands on an academic scientist's time have increased. Thus, it is increasingly common to see smaller and smaller audiences at departmental seminars, much to the detriment of the transfer, development, and discussion of scientific ideas. I therefore encourage you not to let this activity drop from your list entirely, even if the advertised seminar does not appear to be related directly to your specific field.[2]

Conferences, especially those that require some travelling to attend, are a different kettle of fish entirely. Despite your best intentions, you will never be as productive while attending a conference as

[2] In fact, I argue that within reason, the more divergent a seminar topic is from your own field, the more you are likely to get out of it. This is because different topics provide a higher likelihood of discovering new ideas and/or approaches that are unknown or only nascent in your own area of research.

you are in the comfort of your own office. Regardless, attending conferences is an essential part of any scientist's life (Chapter 20), and so you need to pick wisely and efficiently which ones to attend. There is no sense at all in attending every possible conference in your area of research, even if you could afford to do so, for you would never produce any science of your own. On the contrary, avoiding all conferences could alienate you from your colleagues and hinder your long-term progress.

PRIORITY 13: EDITING FOR A JOURNAL

I have placed editing articles ahead of reviewing them for the simple reason that editing requires a good degree more organisational capacity than one-manuscript-at-a-time reviewing. If you are not yet an editor for a journal, then this obviously does not apply. However, if you are an editor, or are contemplating becoming one soon, then be prepared for a reasonably demanding schedule that will undoubtedly steal time from other priorities. A role as an editor is important for you career (Chapter 9), but it comes at a cost; for example, I reviewed far fewer manuscripts for other journals once I became an editor.

PRIORITY 14: REVIEWING MANUSCRIPTS FOR JOURNALS

No scientist likes a lengthy review process for her submitted manuscripts, so do not be the lowest common denominator in the process and submit your reviews later than you had promised (Chapter 8). While reviewing should never come before writing your own articles, it is still an essential task that requires some planning.

PRIORITY 15: BLOGGING AND SOCIAL MEDIA

Unless you have skipped ahead and already read Chapter 21, you might not necessarily appreciate the importance of social engagement outside of the standard media. I probably spend too much time blogging and on social media promoting my science, but for the same reasons standard media engagement is important (*Priority 8*), so too is social media. However, I have placed these activities lower on the global list because they should never cut into the more important activities listed above; rather, they should enhance your writing process and skills (Chapter 2), and be used to make positive changes to society (Chapters 23 and 24).

PRIORITY 16: E-MAILING

Has there ever been invented a more simultaneously wonderful and ghastly technology? I fully admit that I loathe e-mailing, yet I am bound to it for academic survival. I could not, for example, write as many scientific articles as I have without this near-instantaneous mode of global communication. At the same time, I mourn the untold hours I have spent responding to irrelevant diatribes from colleagues, administrators, students, and members of the public. I hate e-mail, and I love it. My only advice is therefore to keep your e-mailing curt and to the point, avoid unnecessary and fruitless discussions, and answer only the most essential requests. If you can, pick up the telephone and solve the issue in real time as efficiently as possible. My cleverest and most efficient colleagues tend to e-mail only during specific hours of the day to avoid the temptation of wasting time responding to messages as they trickle into their mailbox.

PRIORITY 17: MEETINGS

I despise these even more than e-mail, believe it or not. But like e-mails, the meeting is a necessary evil, and you can choose to attend only the most essential. I know colleagues who attend any and all meetings for what appears to be the simple pleasure of listening to themselves speak out loud. If you want to be efficient and effective, avoid this temptation no matter how highly you rate yourself on the narcissism scale. Next to e-mail, meetings (especially those of an administrative flavour) are some of the biggest wastes of an academic scientist's time. Committees to which you sign up are slightly different, in that you have already agreed to commit the time required.

While the number of committees on which you sit can be viewed favourably by your administrators, choose which ones to attend carefully and sparingly.

PRIORITY 18: WRITING RECOMMENDATION LETTERS

If you are not yet far into your scientific career, you might be forgiven for thinking this to be far from an onerous task. However, I challenge you to recall the number of times you have asked your mentors to provide letters of support for scholarship, job, prize, and grant applications. Any mentor who has been at the science pursuit for some time would likely tell a sad tale of the hundreds of recommendation letters written over her career, should you ever be bothered to ask. Clearly this should be lower on the time-priority list, but as for all other tasks, you at least owe it to your charges for the simple reason that others wrote them for you. A useful trick to minimise the time required to write these is to ask the requestor to provide you with a draft letter of the components he would like to see covered, to which you can then add your flair, turn of phrase, and letterhead.

PRIORITY 19: CASUAL ASSESSMENT OF A PEER'S WORK

I have to fight the emotion of anger or annoyance at the impudence of some of my peers who casually and unashamedly ask me to provide comment on their draft manuscript or grant proposal, as if I had all the time in the world to offer them my insights. Do not misinterpret me, for it is a generous and noble thing to do, especially if you can help out the next generation of scientists to negotiate their way through the maze of a scientific career. However, if you cannot clearly identify a personal benefit in the request (i.e., no money or co-authorships are on the table), then provide this service only if and when time permits. For those requesting the assistance, contrite and suitably polite language can go a long way to getting the critical feedback you want.

PRIORITY 20: ADMINISTRATIVE REPORTING

Periodic reviews of your progress, lists of publications, student performance indicators, and myriad other ways university administrators
love to quantify their charges' performance are also part of the academic scientist's territory. While they cannot always be avoided, I suggest that the personal 'reporting' mechanism you use to market your
scientific results (Chapter 21) can also double as a pre-prepared performance indicator for your administrative masters. In other words,
never do a report twice.

17 Work–Life Balance

I could start and end this chapter with a single sentence – *a work–life balance in academia does not exist*. But like most things, it is unequivocally more complex than that slightly patronising absolute. No, there will never be a point in your career when you will say to yourself: 'This is it – I have achieved balance between my career and my personal life', but there are things one can do to make chasing that ideal a little less chaotic. Neither is the challenge of trying to achieve a healthy work–life balance in any way particular to academia; I suppose any demanding career would present similar obstacles. In some ways, academia can even be much more forgiving than other highly demanding careers in this regard. Nor is there a one-size-fits-all formula to help you along the way – each scientist will have both professional and personal circumstances that require their own special blend of organisation and sacrifices.

Apart from the odd scientist who chooses to live single and childless, most scientists are people with otherwise normal family lives (despite public opinion to the contrary). Most scientists have a partner, and many reproduce and have children. But despite some successful scientist couples,[1] most scientists' families are not made up of other scientists, meaning that there is nearly always an inherent friction between scientific and family duties. Unless you are an utterly neglectful parent or partner, or an ineffectual scientist who spends all his time out of the office and with his family, this means that most scientists are faced with having to make sacrifices regarding

---

[1] Couples comprising two scientists are probably more common than you might think, but even they too have sacrifices to make, especially if children are involved.

their careers for the sake of their family, and *vice versa*. And I suppose this is really the second lesson of this chapter – that sacrifices are necessary and so one must accept that they will have to be made. I tend to view my own career as lurching from one crisis to the next, and I am now finally fine with that. Once you have scaled that psychological barrier, the rest is much easier.

If you entered the science arena expecting to work from 09:00 to 17:00, Monday to Friday, you will be profoundly disappointed. Science is not that sort of a job. I hesitate to prescribe a particular number of hours per week that an effective scientist should work because it varies so widely, but it is decidedly more than 40 hours per week.[2] As I mentioned in Chapter 16, different times of the year and activities will require more or less than that (e.g., grant-writing season, field work, intensive laboratory experimentation, etc.), but get used to waking up before the songbirds do, or going to bed well after everyone in your house is long since asleep.

Yet this demanding job is possibly one of the most flexible of the high-profile careers one can choose. I dare say that scientists in general work about as much as medical practitioners or lawyers (or more), yet they can choose *when* and *where* to put in those hours in a way that the others could only dream of doing. The requirement of being around the office for set hours of lecturing or specific administrative meetings notwithstanding, the academic scientist has the near-unique opportunity to customise her agenda to suit all the demands on her time. And this power of flexibility only increases the more you practice the guidelines set out in this book, for academia is unparalleled in the opportunities it presents *provided* that you are productive in the areas that your peers deem important – the academic 'bottom line' of publishing papers, obtaining grants, and teaching courses. In

---

[2] Throughout most of my career I can confidently, yet conservatively estimate I worked an average of 80 hours a week, although I wish to reiterate that this is entirely an individual choice and not necessarily a requirement. If you are more efficient than I am, yours could be substantially less.

reality, if you are consistently achieving that bottom line, then the number of hours you work, and when and where you work them, become largely irrelevant.

Early-career scientists in particular should rejoice that technology has never been more conducive to flexibility in work patterns. Internet access, even in many developing nations, is nearly ubiquitous, and e-mail means that you can collaborate with other scientists from around the world easily and fluidly. Even video and desktop-sharing software now allows you to attend meetings virtually, and with anyone from practically anywhere in the world from the comfort of your own home.[3] So, take advantage of that flexibility and work from wherever and whenever you can. If that means at the kitchen

---

[3] Although I recommend putting on your trousers first.

table drinking coffee before the kids wake up, in the car waiting for them to finish a music lesson, on the train to the office, or lying on your bed in the middle of the day reading a paper, then so be it. Whatever pattern works for you, I urge you not to go into the office every single day like clockwork.

Having children is perhaps the most disruptive thing you can do to a scientific career, but it need not be a crisis or a time of low productivity. Many different factors will contribute to the ease with which you balance reproduction with you career. One of the simplest bits of advice I have is that if you have a choice, perhaps consider putting off having children for as long as is possible. There are of course both advantages and disadvantages to delaying your reproduction, just as there are (different ones) for having children early. However, if you have the opportunity to build your track record without the time commitment, cost, and distractions of rearing children, then you just might find that you are in a more comfortable, financially secure, and established situation that will make the task more compatible with your scientific endeavours. Of course, there is nothing 'wrong' whatsoever with having children early, but having a lot of money, an extremely supportive extended family, a supportive partner, and preferably all of these combined, will make life for you a lot easier when the children eventually do arrive. Money allows you to hire help when you so desperately need it, family members can take on some of your parental duties, and having an understanding, selfless partner willing to sacrifice relatively more than you in this regard is a precious thing indeed. For the sake of the planet and your own sanity, you might also like to consider limiting the number of children you produce – the investment required grows exponentially with each additional mouth to feed.

Some work environments make the career–children balancing act a lot easier than others, but it really does depend on your individual situation. If you do happen to find yourself employed by an institution that is making your personal life difficult, such as not being flexible with respect to working hours, providing insufficient leave opportunities, or even discouraging you from enjoying a healthy family life, then you can certainly consider moving to an employer who offers a better package. Maternity and paternity leave, especially during the early years of child rearing, are essential, so consider negotiating those into your contract when offered a position, and certainly be mindful of this aspect in your own charges once you progress to the rank of 'employer' (e.g., as a lab head). Also consider opportunities to mix family and work, such as occasionally taking children into the office (provided your lab mates are child-friendly), or travelling with them to conferences. Even field work can be done with small children in tow, especially if you have a little help.

I could try here to wade into the tempestuous waters of gender issues in the pursuit of the work–life balance, especially with regard to children, but I cannot even convince myself that I can (nor should) offer anything useful to women in particular. As I discussed in Chapter 15, even science is stacked against women, so I will refrain from offering you some half-cocked, 'mansplained' advice.[4] If you are a female scientist seeking gender-specific advice, I can, however, suggest some excellent material to consult (89–91); better yet, I can recommend that you speak to established female scientists to garner their perspectives. If you are a male scientist and in any position of power, you would also be wise to consider that in general it is much easier for you than it is for women, and take this into account in your day-to-day actions and behaviour.

The crux of being an effective scientist with a happy, healthy family is being extremely organised. Even if you are in the enviable position of not having to worry about money, you will still have to be superlatively prepared. I admit that I am not the most organised chap in science, but my partner is, and that makes all the difference at home. Regardless, you will be in a near-constant state of planning with your partner and perhaps extended family about who takes

---

[4] For those unfamiliar with the term, 'mansplaining' is when a man attempts to oversimplify a concept in a condescending, patronising, overconfident, and often inaccurate manner when talking to a woman.

whom to where for such-and-such activity, who will do the shopping, who will do the cooking and when, who will clean what, etc. I also know many scientists who have particular 'non-negotiables' when it comes to family and work. For example, some will never miss an evening meal with their family, and others will always be there for school drop-offs and pick-ups. Once your employer, students, and colleagues are familiar with your non-negotiables, there is nearly never a problem. Even when conflicts do arise, negotiations with your work mates are possible. I even know of one extremely successful physicist who negotiated trading babysitting from her PhD students in exchange for reading their thesis chapters in a timely fashion, and I myself have had students house-sit for me while I have been overseas. The key here is to consider all the options and not to fear asking for a little assistance.

Finally, I can confidently say that while the pursuit of a work–life balance can be frustrating, fatiguing, and stressful, it is absolutely something to which every scientist must aspire. Just like adequate exercise, a healthy diet, and minimal substance abuse will all contribute to enhancing your cognitive capacity, so too will a happy family life provide you with the mental stability and positive outlook to do your job well. It is just a fact – healthy, happy, and loved scientists do better science.

18　Managing Stress

As professions go, being a scientist probably does not top the list for most stressful, although if you are drilling ice cores in Greenland, photographing deep-sea life from submersibles, setting up seismography equipment on the slopes of active volcanoes, diving with predatory sharks, or scaling 75-metre trees in the Amazon rain forest to collect beetles, then stress is just part of the job. However, I am not going to discuss that kind of stress; rather, I am referring to the day-to-day stress of a demanding academic environment.

If you are a PhD student at the moment and preparing for a life in academia, then you will have at least an inkling of the kind of stresses scientists experience. Even the stress arising from learning how to be self-motivated enough to write an entire thesis, let alone the stresses associated with learning the literature, living on the scent of an oily rag, and a habitual lack of sleep, are reasonable indicators of what is to come. While writing a thesis is unquestionably stressful, it is also ironically the time of your scientific career when you experience the least amount of chronic stress. Acute though PhD pressure can be, it is fleeting and unidimensional,[1] whereas the stress of the career scientist is insidious and multifaceted. The cumulative stress of academia grows as one progresses from being a student, through postdoctoral life, to early-career lectureship, and all the way to tenured professorship. Will I be awarded that grant? Will the editors accept my manuscript? Will I be promoted? How long will I have a job? How do I make sure my lab members succeed? Will I be invited to that conference? Do my peers respect me? How do I recover from

---

[1] In the sense of having to be focussed on a single thing – the PhD itself.

that critique of my research? If you do not learn how to deal with these stresses along the way, you are likely setting yourself up for a big crisis somewhere down the track. Effective stress management is at least partially circumvented by following some of the guidelines discussed in previous chapters, but those by themselves are still insufficient to avoid cracking. I am not a psychologist, so I cannot profess to offer generic recommendations for stress avoidance – there are plenty of published books and expert therapists that cover the topic more comprehensively. Instead, I will provide some tips that my colleagues and I have found to be useful in that regard.

In the day-to-day routine of being a scientist, one activity in particular is simultaneously a blessing and a curse – e-mail. I discussed this little devil in Chapter 16, but it is worth reiterating here that e-mail – rather, the messages delivered by it – can be an immense source of stress. There is the stress associated with pressure to respond quickly to urgent requests (from supervisors, administrators, collaborators, granting agencies, editors, and so on), the stress arising from e-mails that you really should have responded to weeks ago, but still have not yet, and stress from messages that are nasty, vindictive, or even libel received from angry colleagues or misinformed members of the public. Regarding that last source, some scientists might find themselves on the receiving end of political dismay that can turn rather ugly indeed. If you investigate anything to do with climate change, you can expect to receive hate mail from time to time. If you challenge industry or any political decisions based on your research, then expect to receive even more. There are, in fact, myriad ways to whip up factions of the public based on comments you might make in the media, or even casual remarks on social media (see Chapter 21). In today's world of hyper-connectivity, anyone can search online for your e-mail address and fill your inbox with vitriol or even threats of physical abuse. Hopefully most of you will never experience such directed hatred, but if you do, I can recommend that you collate any hate mail into a dedicated folder in your e-mail software. The reason for this is to keep evidence if ever the police need to get involved (threats), or to amuse yourself later once you have calmed down a bit.[2]

Hate mail aside, a good way to minimise stress associated with an overflowing and unanswered inbox is to choose your time for e-mailing carefully, and stick to prescribed times for this activity. Being constantly harassed by the 'ping' of a new message appearing in your inbox will only exacerbate your total stress load. Having dedicated times set aside to respond (for example, first thing in the morning, during a coffee break or lunch, or just before leaving the office for the day) not only helps alleviate some of that stress, it makes your overall time management more efficient. And for sanity's sake, set up an out-of-office automatic responder if you are travelling. You will be

---

[2] Richard Dawkins, the famous evolutionary biologist and author, and notorious attractant of hate mail from right-wing and religious types, read aloud some of the choicer morsels of his hate mail in an amusing video you can view at YouTube.com. I imagine that it was highly therapeutic for him, but probably not so much for his detractors.

surprised just how few e-mails require an urgent response once the sender knows that you are likely to be temporarily out of reach.

But once you become stressed (and you will), how should you deal with it? One of the first positive moves you can make is to limit your drug use. I am not (necessarily) referring to crystal methamphetamines, cocaine, or even marijuana, but even excessive alcohol and caffeine can be bad for maintaining a healthy psychological state in academia. Most scientists I know are not teetotallers; in fact, they are quite the opposite and therefore not opposed to quaffing far too many beverages during the week. I cannot recall how many times I have figuratively limped home after a five-day conference comprising relentless pub sessions and swear off the grog for a month. As an Australian academic, I am doubly cursed. Copious coffee[3], and dare I say it, caffeinated fizzy drinks,[4] are also mainstays of the scientist who stays up far too late most evenings trying to finish grant applications, reviews, or revisions of their own manuscripts. This is not a book about finding a pathway to good health, but eating well and taking stimulants and depressants sparingly is a good start. Good eating habits are also conducive to good mental performance, so avoid subsisting on junk food and a lack of vegetables.

Similarly, a good exercise routine is a must. There is copious evidence to support the link between improved physical fitness and higher cognitive function (92–98), not to mention the benefits of physical activity for good mental health (92, 94, 99–103). Unfortunately, a busy academic life often dictates that certain extra-curricular activities must be sacrificed, and a regimented exercise routine is too often the victim of the academic lifestyle. From personal experience, I know that even a brisk, 15-minute walk can help me to solve a momentarily

---

[3] I love good coffee, but one tactic I have discovered to reduce my caffeine intake is to limit myself to one strong coffee in the morning, followed by Chinese teas throughout the rest of the afternoon (tiguanyin 铁观音 and pu'erh 普洱茶 are some of my favourites).

[4] Sugar-loaded fizzy drinks are horrible enough for one's health, but happily few scientists these days smoke cigarettes. If you do and claim to be a rationale scientist and you still smoke, then you are clearly deluded.

frustrating analytical problem, or allow me to figure out why a particular line of code is not working. While you might be an active, young scientist now with plenty of exercise built into your daily life or even field work, as you age and the demands of academia become heavier, you will discover that exercise opportunities begin to dwindle. My advice is to find a physical activity you enjoy that can fit into your schedule and do it regularly throughout the week. Weekend-only exercise is insufficient.

I love sleeping, but I am not terribly well disciplined at starting the process; I am what you might call a 'night person'. Many of my close colleagues are precisely the opposite and like to wake before the sun has graced the horizon to get a start on their daily duties. Whatever type of person you are, try to avoid the habit you might have developed as a student of going days on end with insufficient sleep – like with a lack of physical exercise, you will operate at a higher cognitive capacity if your body gets enough sleep. Of course, there will be times when sleep is not forthcoming (like during a conference; see Chapter 20), but do your best to catch up as soon as possible.

You might be thinking to yourself that these are rather intuitive recommendations that everyone ought to know already, and I concur. It does alarm me though just how otherwise extremely intelligent people often neglect themselves and behave in a manner that treats the body as a rubbish tip rather than a temple. The next time you attend a conference, have a good look around you at the attendees and you will understand what I mean. By all means, have fun and be sociable, but neglecting your health is not the best path to follow in becoming an effective scientist. But there is so much more to academic stress management than just living healthily, and had I appreciated them earlier in my career, I could have saved myself much grief and possibly extended my lifespan.[5]

The following recommendations might sound a bit 'New Age', but I can personally attest to their effectiveness, and modern science

---

[5] Although I do not believe that I am about to shuffle off this mortal coil just yet.

backs up my perception. The first of these is meditation, or 'mind-fulness', which is essentially the practice of clearing your head of dis-tracting thoughts and giving your brain time to settle. I used to believe that meditation was for drugged-out hippies and astrologists[6] (clearly non-scientist-like pursuits), but since I have come to realise that med-itation is a surprisingly effective method to alleviate stress and all the bad things it causes. Meditation does not (necessarily) require that you sit in a circle with other people with crossed legs while chant-ing 'ohm'; in fact, it is best practiced alone. Nor does it require much time – you can get away with ten minutes of meditation per day and still reap some of its rewards. How does one meditate? I learned by following a few smartphone apps designed for the busy professional, although there are many courses, books, and online resources avail-able. It also takes a lot of practice to quiet a mind, even for a few minutes, so doing it often and regularly is a good idea.

At risk of sounding even less empirical, the other suggestion is to take up yoga. Yes, yoga. Had I discovered what yoga was capable of doing to my mind when I was younger, I would have been practis-ing it since my undergraduate days at least. There are many forms of yoga, for example from the more physically active *Ashtanga*, to the more meditative *Iyengar*, but all forms essentially help you to keep your mind in shape while you do good things for your body. You can also kill two evil birds with one stone by making yoga your princi-pal physical activity if time does not permit you to do more than one non-academic pursuit.

Finally, make absolutely sure that your office environment is conducive to peace and harmony. Good lighting, climate control, ergonomic office furniture, some semblance of privacy,[7] easy access to kitchen facilities (for hot drinks, heating your lunch, etc.), and com-fortable meeting areas are all elements of a good working environ-ment to promote efficiency, good time management, and a general

---

[6] Believe me, I am not the sort of person who embraces these kinds of behaviours without a clear, empirical demonstration of their effectiveness. I am a born sceptic.

[7] For this reason, I do not support the 'open plan' office layout.

feeling of belonging (Chapters 13 and 14). If you do not enjoy going into the office each day, then there is something wrong with where and in what conditions you do your day-to-day scientific activities. Work out what the problems are, and attempt to fix them. The stress of foreboding before you even sit down at your desk is not the best way to start your day.

PART IV   The Fun Stuff

19 Give Good Talk

As a science student, I will wager that you were obliged to give more than a few presentations in front of your classmates well before you ever wrote your first real scientific manuscript. Presenting our research to our peers is a core skill for any scientist, yet is remarkable just how awful most of us are at presenting our research in an engaging and informative way. We tend to lack any sort of showmanship, and we regularly break the rules of engagement for keeping our audiences interested. Take the average scientist out of the normal conference circuit and ask them to present to the general public, and the situation is far worse.

So, the onus is on you to improve your presentation skills constantly. There are of course many different styles, formats, media, and audiences for scientific presentations, so I will only outline general issues of which you should be aware when preparing and delivering that often stress-inducing presentation. There are also many different durations of a scientific talk – including, but not limited to, everything from a five-minute speed talk to a full-on, dazzling, TED-like[1] performance that can last for over an hour. Many of these principles apply to the full gamut of talk types, although various elements will have to be adjusted according to format. Only experience and plenty of advice can assist you in the process of tweaking.

You should remember that a presentation is a story, not a verbal rendition of a scientific article. I have in fact attended presentations where the uninspired scientist was actually reading snippets from a paper without so much as looking at the audience – *do not do this*. Science is a *method*, but the *results* it unearths are a *story*. Rarely, with

---

[1] TED (Technology, Education and Design) Talks (ted.com).

the exception of talks about methods, should presentations go into much detail at all about how experiments were designed or what sampling protocols were used. You generally will not have enough time to go into this methodological detail. Despite having spent the last few years doing nothing else but these repetitive and menial tasks, they are usually the most boring part of a scientific paper anyway, let alone a verbal description. Think 'Once upon a time, . . . ' and go from there.

Much like preparing to write a scientific paper (Chapter 4), it is important first to get your main message straight in your own mind. Here, the 'Three-Floors Lift Rule' applies, which essentially implies that you should be able to explain (*i*) what you do, (*ii*) what kind of problem you are attempting to solve, and (*iii*) why the whole endeavour is important, to a complete stranger in the time it takes to travel three floors in a lift.[2] I recommend spending considerable time crafting this message, then giving it at random to people you meet. Your

---

[2] 'Elevator' for the North Americans amongst you.

ideal response from these Guinea pigs is for them to open their eyes wide and say 'Wow!', instead of glazing over and looking as though they wish instead that they were being poked with a sharp stick.

PRESENT LIKE A PRO

Following the honing of your message, the general structure of your talk should look something like this:

(a) This is my main message.
(b) This is how I am going to tell you about it.
(c) This is what I told you.

Everything else is detail. Repetition itself is not a bad thing, because people have short attention spans and without constant reminding, the members of the audience start to think about the next talk that they might want to see, the snack that they hope they eat at the break, or the fact that they have not yet prepared their own talk.

I also like stories that are at least a little bit funny. Dry, mono-tonic, emotionless, expressionless, or humourless talks do not just make me want to go to sleep, they make me wish something nasty would happen to the presenter for wasting my time. To avoid people thinking nasty things about you, try to pepper your presentation with cartoons, humour-filled stories, funny anecdotes, and witticisms. Do these things and you will have the audience eating out of your hand. A little humour can also go a long way to distracting those predators out there who might be set on exposing your weaknesses and flaws during question time.

If you are using a software program to create slides, whether it is the standard Microsoft® Powerpoint package, or something flashier like Prezi (prezi.com), CustomShow (customshow.com), or Canva (canva.com), resist the temptation to fill your slides with text. Text is your enemy. I would die a happy man if I never again saw another word on a slide, let alone the surfeit of full sentences I usually see splattered across slides in nearly every conference presentation. Instead, rely mostly on imagery, and if you have to print words at all, be exceedingly brief. Use symbols; avoid articles and verbs; and never compose anything that can be read like a book on screen. At the most, a few words can help guide your narrative and focus the eye of the beholder, but images are always better.

Figures, graphs, and schematics are also wonderful substitutes for text. Figures say a thousand words, whereas a thousand words make you want to slit your own wrists. Maps are wonderful too. If you have anything spatial at all, embed a cool map into your presentation, for geeks love maps, and we are all at least a little bit geeky. Unless you are attending a statistics or mathematics conference, it is a good idea to avoid equations about as much as text. Even I, a self-proclaimed quantitative type, hate seeing them in a presentation because I never have enough time to dissect them and understand what every element means (I do not have a photographic memory nor I am as mathematically gifted as I would like to be). Unless it concerns a particularly simple equation that most people already know, and it is essential to your narrative, avoid them like the proverbial plague. Quotes, on the other hand, are an entirely different matter. Most people love clever, witty quotes from really smart, famous people who sum up a complex concept in a mere sentence. Every discipline has great people who are famous for their wonderful quotes. Use them often (and attribute them accordingly, of course).

This is fairly minor, but it cheeses me right off when I see it done improperly. Please, please, please, reference your slides so that we can find the citations to which you refer. Nor is it sufficient just to put a microscopic 'Bradshaw et al. 1845' at the bottom of the slide and expect anyone to be able to find the original paper. At the very least, give the minimum required detail (e.g., Bradshaw et al. 1845 *J Clever Sci* 15:342) that includes the first author, the year, the (abbreviated) journal title, the volume and the first page number. Anything less is kind of useless, and just a little bit patronising.

Once you have the required number of slides for your talk's prescribed duration – a good rule of thumb here is no more than one slide per minute of intended presentation time – practice, practice, and re-practice giving your talk before you actually give it. If you are away from home and in a hotel getting ready to go to the conference venue, practice five times in front of the bathroom mirror before even thinking about giving it in real time. Another bit of advice that can assist is to practice giving your presentation in front of a mate/spouse/partner/relative once you have your storyline firmly wedged into your brain. Their feedback can be invaluable, as well as often rather humbling. Then give the talk again to the wall; not only is this a great way to ensure that you stay within your time limit (everyone hates a chronologically greedy scientist), you will know that you have a lucid, on-time, and engaging talk once the words just roll off the tongue without much thought. Like lines memorised by actors for a play, so too should your 'lines' be more or less committed to memory.

If you happen to be the shy, introverted type most of the time, or you are just plainly terrorised by the mere thought of getting up in front of a large crowd of people and trying to sound intelligent and scientist-like, then the following advice is particularly pertinent. Being well-prepared notwithstanding, the best way I have found to overcome the butterflies is to remember one simple rule: you are the smartest person in the room. I am not trying to be facetious or deliberately provocative here; it is in fact true. Even if you are a beginner student and find yourself in a room populated with crusty old professors who have spent their entire lives researching your topic, I guarantee that you know more about your project than anyone else, for the main reason that no one else in the world has the time to dedicate all their focus to your specific research. You are the expert, so take some solace in the fact that you know more about your talk than anyone else in the room.

Instead of running out of the room to avoid the throngs of admirers, or (more likely) those who want to rip you a new one, embrace question time. Love it; need it; anticipate it. For nothing says: 'I could not care less about your research' more than no one asking any questions. Of course, make sure that you leave enough time for questions, as well as recalling my explanation that you are the cleverest person in the room, and you will have no problems fielding even the stickiest or most aggressive of questions.

I have a few other pointers to help you during question time: (*i*) *Listen* to the question, *pause, think* about it, and only then *answer* – defensive answers rifled off even before the question has been delivered in its entirety are a sure sign that will sound like an idiot or just full of yourself. (*ii*) Avoid appearing defensive; remember that you are not on trial and that you are not expected to defend your approach. Of course, you might get a question or two about why you used a particular method or apparatus, which you can easily justify. You might even be challenged about your choice, but you can defer any debates about this to take place at the pub afterwards instead of engaging in an on-stage battle in front of your peers. (*iii*) Humour self-centred people who just want to disguise their own diatribe as a 'question'. Do not belittle them in public, no matter how easy it might be or how much you dislike them. Just smile, say something polite, and move on; (*iv*) It is entirely acceptable to say 'I do not know' if you really do not. Nothing screams 'bullshit!' more than someone trying to answer a question for which they clearly have no reasonable answer. Most importantly, stay in the room after your talk/session so that those who did not get a chance to ask their questions, or those who are genuinely interested in your research, have a chance to talk to you. If you can (and you should), go for a beer[3] with them afterwards and hash out a good conversation. Who knows? It could lead to a wonderful, new collaboration.

Finally, on the topic of *what* to present, I offer this little bit of advice. This suggestion might seem to be largely out of your control if, for example, you are a student working on a specific topic. However, I am adamant that you do have an element of choice here. If you are a student and your thesis has, for example, several components with

---

[3] Or your beverage of choice.

potentially global implications, focus on those and do not forget to frame your work within the bigger picture. In other words, do not just explain your results within the context of some largely parochial framework. Controversies are always great, because everyone likes to watch a good fight, and anything involving public or political discord (think of climate change, human rights, financial inequality, et al.) will even keep those at the back wide awake, even in a hot, dark room.

THE JOB-INTERVIEW SEMINAR

Now that you have the necessary information to master your presentation skills, it is prudent to discuss the particulars of giving the best-possible seminar when you apply for a new position. If you have not yet had the opportunity to be grilled (interviewed) for a new job, you might not appreciate the importance of giving the best seminar of your life to increase your chances of getting the job you want. Normally in most academic settings, a group of the most qualified candidates for an advertised position will be invited to give a seminar to the main group (department, school, or centre) for whom they could be eventually working if successful. While all of the previous advice applies to these seminars too, there are some specific issues that the candidate must also ideally take into consideration. I myself have both given many such seminars, and I have sat through many more from prospective candidates in my department. Unfortunately, many of these seminars are just awful, serving only to bathe the aspirants in a most unflattering spotlight of incompetence.

To avoid people like me thinking that people like you are not up to the job on offer, here are some important points to remember. First, it is entirely reasonable to be more nervous for this sort of presentation than your average conference or public seminar; after all, your very salary could depend on it. Practice of course makes perfect (or at least, acceptable), but do note that the audience will be sympathetic to your butterflies. That said, avoid at all costs stuttering through an ill-prepared component of your presentation. Some people use pocket cards with notes, and others rely on presenter notes in software like

Prezi, Powerpoint, or Canva, to get through the sticky spots, although I have it on good authority from several academics employed at some of the world's top-ranked universities that constant note-checking is not a good look. For that reason, memorising and practising your talk many more times than you would do normally are essential – close colleagues and friends make excellent pre-interview test audiences.

Start by telling your audience (which will include members of the selection committee, and quite possibly many more senior representatives of the administration) what your take-home message is, and why your research is important. In other words, do not wait until you are 25 minutes into your presentation before you do this. It is also a good idea to structure your presentation into approximately ten-minute segments that highlight different components of your research portfolio. If, for example, one segment ends up taking you longer than expected, or by reading your audience you realise that they are more interested in one segment relative to another, you can essentially customise your presentation on the fly. This will allow you to demonstrate three important things: that you (*i*) have a diverse and interesting research programme, (*ii*) are versatile, and (*iii*) can keep to time. Depending on the time allotted (usually 45–60 minutes), you can then ensure that you leave adequate time for questions, and your committee will absolutely demand that you leave enough time for questions (after all, this is part of the interview process). You can sink your entire pitch merely by going over time and preventing questions, and more importantly, by delaying your audience from getting a drink at the pub afterwards.

Technical content requiring description is certainly acceptable, but you must remember that you are presenting to a generalist audience who will not necessarily be familiar with the enzymatic steps in the blood-clotting cascade, or the mathematical derivation of specific thermodynamic laws. If you must be technical, explain the elements in gory, simplified detail if you deem them to be essential components of your research portfolio. And just like you started the seminar, repeat your main message at the end, especially the part about why it is so bloody important. One final reminder about those nasty, stumping questions – you are much more likely to be asked questions of this sort because your committee will be testing your resolve and composure. If you are asked something to which you honestly do not know the answer, then it is even more important than usual not to invent one. If you do not have a clue, do not say that you do by bumbling through an embarrassingly vague attempt at a pointless, and likely incorrect, response.

20 Getting the Most Out of Conferences

Conferences, congresses, meetings, and colloquia – whatever your preferred nomenclature – are a mixed bag for scientists, and their role in influencing your effectiveness will change over the course of your career. However, some might argue today that the scientific conference is a thing of the past. Now with super-fast internet, social media, open-access publication, Skype and her congeners, and slick video-conferencing capability, why would we need to travel halfway around the planet to sit day after day in some stuffy room to listen to some boring snippets of half-finished research? Why indeed would you choose to experience mind- and body-numbing jet lag, waste a sizeable proportion of your research budget, and emit a bunker-load of greenhouse gases, just to listen to some boring old farts tell you the same thing they have been doing for the last 20 years? If that were all conferences had to offer (or take away), you would probably never attend another for the rest of your career. Thankfully, that is not why we go.

It is indeed true that it is much easier just to go online these days and see what your colleagues are up to (if they do not have a good online profile, that is their own undoing – see Chapter 21). It is also becoming exceedingly easy to download important papers from almost anyone and from almost any journal.[1] So clearly, conferences are no longer primarily about finding out what research is being done, even if they once were. In fact, I argue that it was really never their role at all.

---

[1] Although see Chapter 8 for a discussion of the profiteering behaviour of most scientific publishing companies.

The most important parts of a conference are in fact the social events. That might sound like a bit of joke, but I am genuinely serious. Although conference social events can often be little more than a piss-up for intellectuals, or (mainly for the younger scientists) an opportunity to woo attractive, like-minded people,[2] the conference social events are probably the best place to gauge future collaborative opportunities.

I still vividly remember my first scientific conference. I was a mere pup, having not even completed the first year of my Masters degree. In retrospect, it was an exceedingly colloquial and specialised conference attended by few noteworthy scientists. But at the time it seemed as though I had stepped into an alternate universe of hugely important people all possessing brains the size of a planet. Without any exaggeration, I was completely overwhelmed and more than just a little bit intimidated. But to this day I do not remember a single presentation I sat through, or even a snippet of new knowledge I gained; rather, I remember a few faces, a few laughs, and a general spirit of discovery.

There is nothing more disarming than sitting at a table with a well-lubricated potential colleague in a pub, club, restaurant, or banquet hall and discussing the finer points of scientific theory. Usually there are also, of course, healthy doses of gossip, rumour, and scuttlebutt to keep the conversation interesting. But in general, I leave most conferences having developed at least a few more relationships with kindred minds with whom I hope to collaborate sometime in the future. Had I not met them outside of the prim and proper confines of the professional conference atmosphere, I might never have known how truly intelligent, and how truly cool, these people are. That says nothing, of course, of the impossibility of determining their scientific coolness over the internet. If you are a student or a young scientist, these opportunities are not just golden, they are essential.

---

[2] Beware the academic romance (Chapter 13).

When else will you get to meet the biggest wig in your field, or have a chance to get to know a famous name? Do not be shy – walk right up, present yourself, and offer to buy them a drink – works every time.

It works both ways, of course. I always say that life is too short to work with reprobates. Chances are, with the encouragement of a little CH_3CH_2OH, potentially bad-news relationships can be averted by discerning the true character of a poisonous collaborator. However, make sure that your own indulgence does not make you the inadvertent bellwether of doom for someone else (especially if that someone else is a good deal farther along in her or his career). Burned bridges take far too much effort and time to repair.

It is therefore essential to try to attend the icebreaker, the poster sessions, the mixers, the banquets, the pub crawls, and any other social event you can buy or steal your way into. You are a teetotaller, perchance? No problem – the advantage is yours, but still attend the events. These days the biggest conferences are difficult to negotiate, and especially with umpteen concurrent sessions. In general, I recommend avoiding the biggest conferences and shooting instead for the magic number of 100–500 attendees. There are exceptions to this, but if a conference becomes too populated it is difficult to get the best social atmosphere, which stuffs up the main reason for being there in the first place. So, do not beat yourself up unnecessarily if you do not make it to all the early-morning sessions.

I also have a few other recommendations:

If I am attending a talk and I notice that the speaker might be interested in one of my papers given the subject matter on display, and she or he clearly did not have the opportunity to read it yet, then by all means, e-mail a copy to her or him as you sit there with your iThing or laptop. This little trick has earned me more than one fruitful collaboration (and possibly a few extra citations).

If you are walking along a corridor and notice a long-lost colleague sitting there by herself, make sure you take the time to sit with her for at least a few minutes. Even if you end up missing that one talk you wanted to see, the social interaction is worth more than being just another slack-jawed face in the crowd.

Get good coffee. From a proper café, because conference-break coffee is universally weak and disgusting[3].

Plan a dinner or something with a few colleagues or co-authors. Often these little *soirées* end up being a marvellous stimulus to re-engage in some inspiring and novel science.

As much as it will probably be more expensive, in a less-scenic part of the city and in a generic megahotel, I generally try to choose accommodation as close to the conference venue as possible (often in the very same hotel). When your room is close, you can nip off for a quick *siesta* or costume change, return promptly to retrieve a forgotten item (like the conference name tag, or the flash drive containing your presentation), or get in that last practice session before your talk (see Chapter 19). Staying too

---

[3] The Italians have a great expression for such coffee: *acqua sporca* ('dirty water').

far away not only removes these possibilities, it increases transit time and ends up being more tiring in the long run. It also removes opportunities for impromptu social events.

One final word on technology and conferences that might give you at least pause to consider how to engage your peers in attendance. It is something I have noticed over the years attending scientific conferences and seminars – the number of questions, and more importantly their quality, have declined. I admit that it is an anecdotal observation, and it might just be that my perspective has changed, but I am convinced that it is a real trend. There are many possible contributing factors, such as increasingly over-attended conferences with multiple concurrent sessions and less time for each of us to present our work. However, I think the main reason is that we now all appear to be glued to our electronic devices.

I am referring to the Twitteratti, of course. I am being just slightly hypocritical here, for I am an avid Twitter [*insert your favourite social medium here*] user, and I regularly use my devices at conferences. It is in fact from my own experience, as well as my observation of other attendees, that I have reached the conclusion that the demise of the question is mainly due to our fascination with devices. I understand why we are fascinated with them, for apart from the sheer joy of receiving yet another meaningless e-mail, at least my device entertains me when the real happenings around me are emphatically more boring.

This is the problem, of course, and why electronic devices put you into an ever-descending vortex of disinterest. I know that when I am busy trying to write the cleverest tweet of the session that I am not actually listening to all the words that come out of the presenter's mouth. I know that I often miss key information that passes too quickly from slide to slide, such that by the end of the talk I am less likely to ask a question because I might merely have missed something that was already clearly explained. No one wants to pose a question when the answer has already been given, or you risk looking like a bit of a loser.

Even in cases where there is sufficient time for questions, there appear to be fewer people willing (or able) to ask them. Even when someone does ask one, they tend to be banal requests for clarification. Long-gone are the days that attendees rip into a presenter during question time, challenging his hypotheses, dissecting his methods, posing philosophical alternatives, or contemplating the greater implications of his work. I am sure few people want to be the focus of such examination, but like I explained in Chapter 7, it is better to receive negative comments than none at all.

Social media is no longer the bastion of just a few technophiles either – live-tweeting is taking over conferences to a degree that perhaps few expected. There are possibly better ways to do things, to which I add that maybe it is time that only a few of us took up the challenge and we appointed dedicated Twitter (or equivalent) presentation chroniclers. In something of an auto-therapeutic gesture, I therefore implore you to reconsider bringing your devices to conference sessions, or at least avoid playing with them excessively during the talks themselves. We might end up having a much more engaging, scientifically meaningful, and fun experience at conferences as a result, the more important social aspects of the conference notwithstanding.

21 Science for the Masses

We can safely argue that nearly all aspects of modern human life owe their existence to science. Electric lights, mass food production, transport, air conditioning, medicine, heating, clothing manufacture, etc. are all the products of scientific research. If we therefore convinced ourselves that our scientific endeavours were merely of interest to other scientists, then we would not only be incorrect, we would be selfish, short-sighted, and historically ignorant. Even the most theoretical and 'blue skies' research can be useful and interesting to non-scientists. We are therefore compelled to extend our science results and their implications to as many people as possible (104). As if we needed more jobs to do and expertise to acquire! Unfortunately, effective public engagement is something that most scientists have done poorly since the advent of modern communication technologies, so mastering a good communication strategy should be something every developing scientist should try to improve.

The five main communication objectives for scientists (104) are:

1. informing the public about science;
2. exciting the public about science;
3. strengthening the public's trust in science;
4. tailoring messages about science; and
5. defending science from misinformation.

I argue strongly that each one of us should try to participate in at least some aspect of each of these, and this chapter is dedicated to making that process a little easier for you.

ONLINE PRESENCE

It astounds me every time I hear about a scientist who is reluctant to place his track record on the internet. Now, it could be argued that I can be a little over-the-top when it comes to my own web presence (some have labelled me a 'media tart', but I do not mind), but I am convinced that without a strong, regularly updated web presence, you are doing yourself a horrible disservice.

Let us go through the regularly raised objections that some academics make for avoiding the investment in a strong web presence: (*i*) my employer will get angry; (*ii*) my track record is not good enough (i.e., I am embarrassed); (*iii*) what I do is no one else's business; (*iv*) I could not be bothered – it is too much work; and (*v*) no one will read it anyway, so what is the point? While there might be some truth to items *i* and *ii* – although the justification is weak or often plainly untrue – the last three are just wrong.

(i) *My Employer will Get Angry*

Yes, some employers (e.g., almost all government departments) are particularly nervous about their employees talking about what they are paid to do (strange, I know), especially when their own scientists contradict what their evidence-free, ideologically motivated politicians-in-charge are saying. There are so many examples of this that I do not need to go into much detail here.[1] No matter what idiot politician happens to be trying to gag 'their' scientists from speaking to the media, I still have never heard of a government agency preventing its employees from saying great things about themselves online. After all, it is the reputation of the institution that is at stake, so it will almost always welcome positive stories. At least a statement of what you do and a list of your achievements should not piss off even the most paranoid employer.

---

[1] But of course for your elitist pleasure, you can snigger at the stupidity, and shake your head in despair, at the actions of people and organisations like American Scott Pruitt and the United States Environmental Protection Agency (EPA), former Canadian Prime Minister Stephen Harper and almost every scientist in the employ of the Canadian Government, the Australian Government and scientists employed at the Commonwealth Scientific and Industrial Research Organisation (CSIRO), and the Chinese Academy of Science (which is *inter alia*, a political unit of the National People's Congress) and any Chinese scientist who ever wanted to be funded.

(ii) My Track Record Is Not Good Enough

While this might be true, especially for the neophytes amongst you, it should never, ever stop you from cobbling together an online profile. In fact, your track record and your internet presence are intimately tied together (see more on *Point iii* below), so a downward spiral develops if you hide yourself away. In other words, your track record is not likely to improve as quickly if you are internet-shy. If you are an early-career researcher, then this perspective is ridiculous; of course you probably do not have a stellar track record yet – it takes time to develop. When I search online for an early-career researcher's profile, I will always take age and experience into account, and so will everyone else.

(iii) It Is No One Else's Business

Wrong. First, most scientific research – perhaps apart from a good deal of medical research – is publicly funded these days. So, you not only have a responsibility to let the people who funded you (the taxpayers) know what you do, you have a duty to do so. Second, if you are that arrogant to think that science is only a personal endeavour, you are possibly contributing to the increase in science illiteracy and denial amongst the general public. Shame on you.

(iv) It Is Just Too Much Work

I beg to differ. Most people at least have some semblance of a list of their publication record tucked away somewhere on their personal computer. If you are an academic, you are in fact obliged these days to report your scientific output to your university, to your funding agencies, and to your research partners. It is really little effort to update a webpage with this basic information. I am not saying that blogging or Tweeting is everyone's cup of tea (nor is it strictly necessary, but see later in this chapter), but at the very least, a quick overview of your projects and publications (with web links) is a bare minimum.

(v)   *No One Will Read it Anyway*

Unless you are a scientist who (a) never publishes, (b) never gets grants, (c) is not looking for work, and/or (d) does not do much at all, then this contention is utterly false. When I am attending a conference and someone I do not know well is presenting, I generally search for their online profile so I can get a better background of the person's research interests. Or if someone e-mails me out of the blue to ask a question or to invite me to some event, I almost always search for their profile for nearly the same reasons.

It seems almost pointless to state, but if you are seeking employment and you do not have an online profile, you will most likely leave a bad impression with the interview committee who will want to ascertain how you present yourself to the public. No online profile means that you are immediately at a disadvantage relative to your competitors. Even with your curriculum vitae and publications list in-hand, as a selection committee member, I will *always* search for your online profile.

There are also many other obvious advantages of having a good and professional online profile. When a journalist requires some expert opinion (see Chapter 22), she generally searches online (just like everyone else these days). When a policy wonk needs some advice, he might do the same thing. In fact, your web presence is THE PRINCIPAL means by which people get to 'know' you – the internet has replaced all other search methods in this regard.

I have a few other pointers to offer you on building your web presence. While your institution's web pages might be cumbersome, out-of-date, and nearly impossible to update,[2] there are many alternatives. I am flabbergasted that so many scientists still have not discovered Google Scholar (scholar.google.com); for the sake of your career, spend ten minutes and sign up to get a Google Scholar profile. There are other web services for this sort of thing, like ResearcherID (researcherid.com), ORCID (orcid.org), Research Gate

---

[2] Although as a general rule, this has been improving slowly.

(researchgate.net), Academia.edu, LinkedIn (linkedin.com), etc., so you have plenty of free services from which to choose. Even Facebook (facebook.com) can act as a 'professional' website if set up with that express objective.

I really need not write this final recommendation, but for those of you who might be travelling a bit slower on the intellectual highway than the rest of us, this is for you. Avoid – at all costs – putting up semi-naked[3] photos of yourself on Facebook, or going on some ill-advised rant on someone else's blog or news site. If most people knew just how easy it was for others to cyber-stalk them, they would not document all their foolishness online. In other words, try to keep all of your publicly accessible web content as professional as possible. Remember, your colleagues are watching you. You can, of course, still have your private profiles, but for your own sake make absolutely sure that they remain private. If you want to be entirely secure, I recommend keeping anything personal entirely out of cyberspace[4] no

---

[3] Or completely naked.

[4] One word – Wikileaks (wikileaks.org).

matter what a company might claim regarding privacy and security. Even e-mail is no longer private[5] if someone demands that your institution release their servers' data, or they are simply hacked (remember ClimateGate?)[6] (105). *Caveat conscriptor.*[7]

SCIENCE BLOGGING

I vividly remember my lukewarm reaction the first time someone suggested that I should blog. This was back in 2008 when science blogging was not yet much of a thing, so I first had to ask (shamefacedly) what a blog was.[8] Once explained, I detested the idea even more, for how could such a busy scientist possibly find the time – let alone the inclination – to write copious amounts of off-the-cuff verbiage for which he would never receive any formal credit? I still do not really know why I proceeded to do so anyway,[9] although a decade and over 700 blog posts later I can confidently say it was one of the best decisions I have yet made in my career.

The main reason scientists should consider blogging is the cold, hard fact that not nearly enough people read our scientific papers. Most scientists are lucky if a few of their papers ever top 100 citations, and I would even wager that most are read by only a handful of specialists over their lifespan (there are exceptions, of course, but these are rare when placed into the context of the entire body of published scientific literature). If you are already a publishing scientist, you might have already experienced the disappointment of realising that the blood, sweat, and tears shed over each and every paper you write is largely for nought considering just how few people will ever

---

[5] If you do regularly need to send sensitive information via e-mail, then you might consider signing up to a secure e-mail service (e.g., ProtonMail.com).

[6] The irreparable damage done to science (and the scientists involved) is exemplified by the 2009 hack of the Climate Research Unit's (CRU) e-mail servers at the University of East Anglia in the United Kingdom.

[7] 'Let the writer beware'.

[8] 'Blog' is short for 'weblog', and is typically a reverse-chronologically ordered series of short (normally between 500 and 2000 words) musings (posts) about almost any subject imaginable.

[9] In reality, I was enthusiastically encouraged by a former Microsoft employee whom my university had hired as a communications consultant. I think they got their money's worth out of his salary.

read about your hard-won results. It is simply too depressing to contemplate, especially considering that the sum of human knowledge is so vast and expanding that this phenomenon will only ever get worse. For those reasons alone, the prospect that blogging about your own work could widen your readership by orders of magnitude is worth considering.

Even your audience diversifies when you blog. From school children to journalists, to politicians to scientists in other disciplines, to people who would never think to read (or be able to understand) your technical papers, all these will suddenly be able to digest your complex research results in at least their basic form. In essence, a blog post becomes your own personal newspaper article about your work. A greater audience means that you will also likely get your work across to more of your scientist peers who will be interested in collaborating with you. Yes, even scientists read other scientists' blog posts. I have been approached many times by previously unacquainted peers with ideas for collaboration based solely on the subject of my blog posts.

I also argue that for the same reason you are morally obliged to have a web presence (see previous section), you also have a moral responsibility to disseminate your results as best you can to the taxpayers who funded you. In my view, believing that you have achieved that dissemination once your paper goes online is self-delusional and unfair to the members of society who funded it. But even if you are a cold-hearted bastard that is not swayed by the moral or goodwill arguments, you can justify blogging by appreciating that it will probably lead to a higher number of citations of your scientific articles themselves. But do not just take it from me, there are now empirical studies emerging that show that the more a paper is visualised outside of academia, the more it is cited within it[10] (106–109). Indeed,

---

[10] The potential circularity of this relationship – that highly promoted papers are intrinsically interesting and therefore garner more academic citations irrespective of their media attention – might weaken the argument somewhat, although my own anecdotal experience suggests that I am more likely to cite a paper if I know about it already through its media promotion.

the meteoric rise of the 'Altmetric' as an indicator of a paper's reach outside of academia[11] heralds a new era in how scientific papers are rated and judged by society. If you have not yet collected your own Altmetrics into a single profile using services like ImpactStory.org, I recommend that you do.

Another great reason to blog is to satisfy your basic need for vengeance. I discuss the complexities and intricacies of the sometimes-fraught relationship between scientists and journalists in Chapter 22, but it is inevitable no matter how careful you are that your scientific results will be utterly butchered (or ignored) by a reporter at some point in the future. While engaging with traditional media (newspapers, television, radio) through reporters, journalists,[12] and editors will always be important for the effective scientist (Chapter 22), blogging allows you to circumvent the uncertainty of the process by 'cutting out the middle man'. Thus, a blog acts as a sort of personal 'newspaper' for which you are both the correspondent and content editor. When the inevitable happens and your excellent research is misinterpreted, cherry-picked, or just misunderstood, you can correct the mistakes and deliver your own, ideal news item through the comfort of your very own blog show.[13] Not only can blog posts represent the newsy end-product, they can also serve as the precursor to a press release (see more on this in Chapter 22). Every time I wish to send out a press release, I e-mail a copy of the scientific article in question and the blog post covering the topic to my university's

---

[11] Although the term is normally applied to the service provided by altmetric.com, any social (e.g., Twitter, Facebook, blogs, etc.) or traditional (news sources) media mentions of an academic paper will contribute to its total Altmetric score. You can easily collate your Altmetrics into a single profile using services like ImpactStory .org.

[12] I explicitly differentiate 'journalists' and 'reporters'. I define the former as someone who does actual research about a news event and presents the information objectively. On the other hand, the latter merely regurgitates whatever is fed to them – a beautifully and elaborately presented meal will always look ugly once it is vomited back onto the plate, even though it is made of exactly the same ingredients.

[13] Some scientists prefer the medium of the 'vlog' ('video log') to do this. Anyone with a smart phone can now interview herself in this manner and post the resulting video online using myriad platforms.

media office. It makes my media officer's job of cobbling together an engaging release much easier and reduces the probability of making translation errors.

Blogging also allows for a certain amount of independence from the shackles of institutional restraint. Like people afraid of inadvertently raising the ire of their employers for posting online material (see previous section), blogging 'independently' of your institution's main web presence gives you some (but not necessarily complete) freedom. Be careful with your new-found liberty to express yourself openly – as I mentioned there is no insurance for saying silly things online once you press the 'publish' button. By all means, be provocative and controversial, but do it politely and with evidence to support your claims. Which brings me to yet another advantage of blogging – the opportunity to express opinion and values. Most scientific journals are not the best fora for expressing these, so if your research has applications or implications for society beyond the evidence you present, a blog can nicely round off the research lifecycle by telling all who will listen how the information could and perhaps should be used.

I devoted the entire first part of this book to writing, for perhaps next to experimentation and/or data collection, this is the one task that will occupy most of your time during your career. Even a brilliant scientist is functionally an idiot if he cannot express his concepts, hypotheses, ideas, and arguments clearly and concisely. Blogging is of course 'writing', and in a style that demands simplicity and flair. This means that you will undoubtedly improve your writing skills for conventional scientific paper writing – as well as your facility to write more fluidly – the more you blog. Writing practice is therefore never a waste of time, so blogging[14] can be viewed as a mechanism to improve one of the main ways in which you will be judged by your peers.

---

[14] Which is why I prefer to blog rather than vlog.

If I still have not convinced you that blogging is a good idea for every scientist, let me come back to one of the most common arguments against doing it – that it takes too much time. For the same reasons why it is easy to debunk the argument that it is all too difficult to find the time to keep one's online profile up-to-date, blogging should not really add much to your daily agenda. If you are spending more than a few hours blogging per week, you are probably cutting into important scientific writing time. Limited like this, it is an easier pill to swallow. Remember too that as you practice and become a more proficient writer, you will need to spend less time redacting your blog posts, and your overall writing capacity will become more efficient. You might even end up enjoying the writing process a whole lot more than you do now, and nothing passes the time more quickly than doing something you enjoy.

SOCIAL MEDIA

I regard social media in the same light as I do firearms – extremely powerful tools that if used incorrectly or abusively can cause tremendous harm. Social media has taken over the world in ways that none of us could have predicted even a decade ago, and the same goes for how it has revolutionised science communication. Social media can now start revolutions, topple governments, elect politicians, and destroy careers and reputations. Scientists should be simultaneously fascinated and horrified by that power, but because we are by definition clever people, we should have the wisdom to wield it to our advantage. I suspect too that because of how rapidly social media has expanded through all facets of society that it will also continue to evolve at a staggering rate, meaning that even as I write these words my advice might be becoming obsolete. Regardless, I will try my best to be as generic as possible about how I recommend scientists should engage the public using social media.

The first thing to appreciate about social media is that all of its components and platforms are designed specifically to grow *networks* of individuals who would otherwise not be connected. Viewed in this manner, it becomes clearer why social media offer a remarkable opportunity to maximise the penetration of your scientific messages beyond your peers, or even amongst laypeople who already happen to be interested in your topic. The second-most important thing to realise is that different social media platforms appeal to different sectors of society, such that you would be wise to exploit as many of these representative networks as possible to propagate your public influence.

Nearly everyone in science uses some form of social media today, whether you are a younger person who has accounts with 20 different social-media platforms, or a crusty old professor nearing retirement age who dabbles in Facebook from time to time to keep in touch with your grandchildren. My advice applies to all of you regardless of your interest and involvement, for with a little tweaking of your normal social-media behaviour, you can turn that time investment into an astonishingly effective communication tool. But beware the temptation here to mix business with pleasure – combining your personal or family interests with your professional interactions on social media is a recipe for disaster, and could even lead to nasty online interactions that threaten your right to a happy personal life.

Possibly the most commonly used social media platform for engaged scientists is Twitter[15] (twitter.com). Twitter, for those still confused about what it does, is a means to send out short messages, often with an associated web link, to a network of 'followers'. If you are a seasoned member of the *Twitterati*, then this basic introduction might be beneath you, although I encourage you to keep reading to learn a few tricks on how to use it to your full scientific advantage. What can one say in one (usually abbreviated) sentence? Obviously, not very much, but you can think of the message as a bit of a headline that when accompanied with a web link, can attract someone to

[15] Although as I mentioned above, this could change at a moment's notice; nonetheless, the principles will be approximately the same regardless of the specific platform.

click through to the more detailed source. For example, if a news story about a particular research article is published online, you can simply copy the headline and the link to the original story. Viewed in this manner, I think you can start to see the possibilities of using Twitter to promote your own work. For instance, if you write a blog post about your own research, you can then send out a 'tweet' that briefly describes the main message (or headline), accompanied by a link to the post itself. Easy.

The looming question is to whom are these 'tweets' sent? Building your network of 'followers' is an important part of the social-media game, because just having an account will not guarantee that anyone actually reads what you might post. So, do not be disappointed after subscribing that you have only a few followers – this number will grow over time if you follow a few simple rules:

Rule #1: *Post frequently*. If you post a message only once every few days, you will not entice many people to follow you. A good frequency is something in the vicinity of at least a few tweets per day. On the other hand, avoid tweeting crazily hundreds of times per day – you will not get anything else done otherwise.

Rule #2: *Stay on subject*. Just like you are best to avoid mixing business with pleasure, tweet about a specific area and keep to that. Whether you happen to be an amateur epicurean, a knitophile, a keen traveller, stamp collector, or a sword-wielding Medievaller,[16] make sure that you keep that side of your life apart from your science. Your science-based Twitter feed should be about your general field of science, and not much else. Of course, you may elect to tweet about socio-political implications of your chosen science topic (see Chapters 23 and 24), but make sure it always has that science link.

Rule #3: *Do not rely too heavily on the 'retweet'*. A 'retweet' is simply the copying of someone else's tweet to your own followers

---

[16] We all know at least one scientist who does this sort of thing.

(with attribution). If you never publish anything 'original', you will struggle to grow a large following.

Rule #4: *Backup all contentions with scientific evidence.* Like blogging, I encourage you to express your opinions and points of view freely, but always point your followers to the scientific evidence supporting your contentions. Remember the old proverb: "Opinions are like ... "[17].

Rule #5: *Be (somewhat) controversial.* Like blogging, a little controversy sparks interest, although being controversial only to garner a response is unwise. Always remember Rule #4.

As a regular scientist who follows all of these rules and posts highly pertinent, interesting, and entertaining tweets, you will still never reach the number of followers that most celebrities or other famous people acquire. That is perfectly acceptable, but with time you can certainly build your audience to at least several hundreds who in turn will have hundreds to thousands of followers, and so on. Your potential influence even with a modest audience of a few hundred therefore potentially reaches thousands indirectly, if not hundreds of thousands. Not bad for a few short messages about your science sent out a few times a day.

Of course, Twitter is not the only social-media platform on the block. There are many more, including Facebook, Instagram, LinkedIn, and so on. I do not wish to dwell too much on the specifics of each, for they all have their advantages, limitations, and quirks. Regardless, the same general rules that you use for Twitter should be applied to each network, especially the one regarding not mixing your science with your personal life. The key here is to attempt to tap into many networks to expand your sphere of potential influence.

---

[17] If you are unfamiliar with this saying, it is 'Opinions are like arseholes. Everyone has one'. A brilliant addendum to this expression came recently from Australian cabaret comedian, Tim Minchin: ' ... but I would add that opinions differ significantly from arseholes, in that yours should be constantly and thoroughly examined'. Wise words for any scientist.

But how can you possibly find the time to waste on such frivolous social engagement? One could easily spend most, if not all of one's time tweeting, retweeting, blogging, Facebooking, etc. – clearly this is not a good strategy to achieving effective science. The answer lies in the fact that as *networks*, it is genuinely simple to link different networks such that you only have to spend a minimal amount of time posting a message that is then distributed automatically amongst your different social-media platforms. For example, you can link accounts such that your tweets appear automatically in your Facebook stream, or your blogs automatically get reposted on your LinkedIn page. The key lies in minimising your effort and maximising the number of people you reach.

On the subject of linkages, a strategy that I have found to be particularly useful is the notion that 'all roads lead to Rome'; in other words, I attempt to link all of my social media activity to my blog, which essentially exposes my research and its role within my field. If I can garner more interest in my blog posts via the networking power of multiple social-media platforms, I can create in a sense a landing site from which I can launch my communications assault on the general public.

22 Dealing with the Media

I want you to try a little exercise. Without thinking too much about it, name three famous scientists with whom you do not work or know personally. I bet there is a fair to good[1] chance that you heard about them through some sort of popular medium, like the television, radio, or even online. In other words, you are just as susceptible to mass media as the rest of humanity, even within the confines of science communication. My little thought experiment therefore serves as a basic demonstration of the role of marketing in science, and I argue that the effective scientist is also one who markets her or his work well.

If you think that all you will have to do to win the respect of the public and the scientific community is produce novel and high-quality research written up in high-impact scientific journals, then think again; you are dreadfully mistaken. This 'build it and they will come' philosophy will only ever bring you frustration upon the realisation that the years of blood, sweat, and tears you spend to produce your research findings ends up in your articles being read by only a handful of specialist scientists from a restricted field. It would be enough to make you ask yourself why you bothered at all in the first place. As a young scientist, I was convinced that good science spoke for itself. But scientific writing, and the complex analyses underlying it, are not eloquent or particularly engaging performers to non-scientific audiences. In addition to all the other expectations of an effective career in science, you must also therefore become something of a promoter.

---

[1] In keeping with my plea for quantification, I estimate that this would represent a probability ranging from 0.60 to 0.90.

The problem is that few scientists have any formal training in media or journalism, and fewer still have ever been taught anything to do with performance. Nor do we ever complain about having too much time on our hands with which we could fill with media training and acting lessons. Before I continue, let me assure you that an effective scientist is not required to become the next David Attenborough, Brian Cox, Neil deGrasse Tyson, or Tim Flannery to make a splash with her science, for these household names are indeed much more performers than practising scientists. However, elements of their flair, enthusiasm, and capacity to engage dilettantes are things to which all scientists should aspire. So how can the often shy, socially awkward, and empirical personality of the average scientist engage the short attention span of the regular citizen who is already overwhelmed by the modern information deluge?

The prospect of speaking out in public might terrify many new scientists,[2] but this is merely a symptom of a lack of experience. Like anything, practice and familiarity will get you there eventually, but the road is long, dusty, and potholed. It would make more sense to be prepared in advance, so this little chapter might serve to make your journey a little more comfortable along the way, and alleviate some of your dread.

PRESS RELEASES

One of the easiest things a marketing-savvy scientist can do to get some attention is to go through the 'official' media channels by producing a 'press release'. Of course, I have already extolled the virtues of being an active participant in social media (Chapter 21), but even the most socially engaged scientist might still fail to gain the attention of the professional media if that is all they were to do. In fact, if you take my advice and start to blog about your research, it is thereafter a simple evolution to the production of a good press release.

---

[2] Placing this chapter in 'The Fun Stuff' part of this book is not an irony lost on me.

What is a 'press release' anyway? I am nearly certain you have seen one before, either on a university's news webpage, or having been e-mailed one from your colleagues. In its most basic form, a science press release is a simple description of the main message of a research project, paper, or report, usually accompanied by some 'quotes' from the main researchers involved. Often the release itself spans no more than a single page, and it is written for the layperson, with as little technical language as possible. A generic example follows:

Media Release
EMBARGOED until XX Somemonth 20XX (paper scheduled for online release at 00:00 on XX Somemonth 20XX)
—

Scientists discover [*an exciting thing*] that will [*change the world forever*]

Scientists at the [*University of Cunning*] have reported the discovery of a [*really important thing*] that could potentially [*improve people's lives in some way*].

Using experimental trials, the research team tested the [*effects of manipulating something*] on the [*phenomenon of something else*].

'This is the first time that we have found evidence that [*this thing*] changes the way that [*that thing*] behaves, and this could have huge implications for the way that we [*do something else*], stated the project's lead researcher, [*Dr Really Cleverscientist*] of the [*University of Cunning*].

[*Professor Generous Labhead*] of the [*University of Brainpower*] added: 'Previously, we had trouble separating the effects of [*one thing*] from [*another thing*] because [*our equipment was too insensitive*]. With [*the invention of a more sophisticated measuring system*], we can now say for certain that [*the one thing*] is the principal reason [*this other thing happens*].

In the team's latest research published today in the [*Journal of Amazing Scientific Results*],[3] [*Dr Cleverscientist*] and her

---

[3] Provide web hyperlink to the published paper.

colleagues found that by [*manipulating this thing*], society will be able [*to achieve something positive it has not yet been able to achieve*], meaning that [*people will be happier and healthier*].

'In other words, we can [*have our cake and eat it too*]', clarified [*Professor Labhead*].

–

The original paper, [*Some really engaging, but technical-sounding paper*] by [*Really Cleverscientist*], [*Colleague 2*], [*Colleague 3*] . . . , and [*Generous Labhead*], will be published online in the [*Journal of Amazing Scientific Results*] at **00:00 (XXT *SomeTimeZone*) or 12:00 (GMT London) on XX *Somemonth* 20XX**.

For more information contact:
Name: [*Dr Really Cleverscientist*], [*Department of A Scientific Discipline*], [*University of Cunning*]
Tel: +XX XX XX XX XX
E-mail: [*really.cleverscient@univ.someplace.cntry*]

Name: [*Media Officer*], Media Office, [*University of Cunning*]
Tel: +YY YY YY YY YY
E-mail: [*media.officer@univ.cunning.cntry*]

My generic little example might seem a bit silly, but it does get across the idea that the average press release is an *extremely* simplified résumé of (mainly[4]) a recently published research article. You would have noticed a complete absence of technical jargon, with no accompanying qualifiers or probability statements. The main result is put right up front, and there is usually a statement about the way in which the world *could* become a better place as a result of the discovery (i.e., the 'application' or 'use'; see also Chapter 23).

There is a certain art to writing a good media release, and I would not expect the average scientist to be able to cobble one

---

[4] It is also quite acceptable to produce a press release without an accompanying research article; other topics include *inter alia* the success of a recent experiment, the announcement of funding, the establishment of a new organisational unit, the publication of a book, or even the addition of a celebrated new member to the lab.

together easily on the first attempt. Typically, that task falls to a dedicated media person within your institution's media, communications, and outreach office (most research institutions will have a variant of such a thing). But do not just send the media officer a PDF of your recently published scientific article and expect her to produce a sharp, witty, and engaging press release that perfectly summarises your main results and their implications. Even the best media officers are not experts in your field, so you will be obliged to help them during the process.

The easiest way to assist your media officer is to send along a link to your recently written or published blog post about the scientific article in question. If you follow the guidelines given in Chapter 21, a good post can be effortlessly transformed into an engaging press release within minutes. Another thing that can speed up the process is to fabricate some 'quotes' that you would like to see appear in the release. These quotes are fabricated in the sense that most likely none of the research team has as yet actually uttered those specific combinations of words. However, it is not a problem to write some clever analogies, metaphors, or witticisms and 'attribute' them to the lead researcher or even a few of the co-authors (although avoid the temptation to give everyone a quote; two people is ideal; three is a maximum). I also tend to send along a 'main message' statement to my media officer to help them understand the principal, over-arching idea the paper conveys.

The press release is typically timed to go out over the main media networks about the same time that the scientific article first appears online. If you have published a scientific article before, you will understand what I mean here; if not, allow me to explain. Many journals – especially the highly ranked ones (see Chapter 6) – impose an 'embargo' on the media until the paper is actually available in its final form online.[5] This means that reporters and journalists are

---

[5] Despite there being no need whatsoever anymore to publish a paper twice, most journals still have an 'online early' version of their papers without an accompanying

not officially allowed to report the paper's findings until that specific date (and more often than not, a specific time of the day on that date). Trusted media outlets can be sent a pre-embargo copy of the press release on the understanding that their reports must not appear beforehand, and this agreement is generally (but not always) respected.

This process highlights another important reason for doing a press release – your media officer will have access to a network of journalistic outlets to which you, the mere mortal scientist, will not. A good media officer will be able to access networks like the *Associated Press* (ap.org), *Reuters* (reuters.com), *Agence-France Presse* (afp.com), amongst many others, that are the life-blood sources of potential stories for thousands of journalists and media outlets across the globe. With a good headline, a well-written release, and a little luck, a syndicated media outlet will pick up your story and spread it around the globe. It is also a good practice to exploit any relationships with journalists with whom you might previously had positive interactions. If you have been interviewed before and found the experience beneficial, it is common for, and indeed, expected of you to contact these people again upon the publication of your next paper. Journalists love a 'scoop', so if you can give them the privilege of primacy on reporting your work,[6] then you can potentially maximise the media coverage of your study. The important thing here is to be proactive beyond merely writing the release itself.

I must stress that writing a press release should in no way be reserved exclusively for those big papers published in the giants

'hard' (paper) copy and their assigned volume and page numbers. Once the specific issue of the journal is published (now more frequently without an actual hard copy printed), then the paper is considered to be in its final form. Some journals even have an 'accepted' form (or abstract) available online prior to the 'online only' version, whereas others (mainly online-only) skip straight to the final format upon acceptance (e.g., the *Public Library of Science* family of journals).

[6] A little plea to journalists here – please, please, please provide the link to the original scientific paper in your news article. It is most frustrating for scientists reading about the recent publication of an interesting study not to have the original source available at our fingertips.

of scientific journals (i.e., the *Nature, Science, Cell,...* kind). Many scientists mistakenly think that only the biggest papers will garner any media interest, when in fact even the smallest and seemingly parochial paper can 'go viral'. I will refrain from giving specific examples from my own research, but I am still surprised that sometimes the most innocuous and seemingly mundane scientific papers that I co-author can make a huge, international media splash. Sometimes, all it seems to take is a great headline, but more often it is probably just down to plain, old-fashioned, good luck. Regardless, it is nearly always a good idea to do a press release for a scientific paper, even if hardly anyone notices. You never know.

THE INTERVIEW

If someone does notice, then there is a good chance that someone will want to interview you – for a newspaper, radio, television, or an online-only news source. Every medium is different, and only experience with all of them will fully prepare you for a snag-free media experience. But most interviews will happen by telephone, and increasingly journalists use computer-based communication applications like *Skype*, *Zoom*, or something similar to interview their 'talent'.[7] If you do not currently use any such communication application, I strongly urge you to change tacks and subscribe to one if you want to be interviewed about your research.

The first step is the initial contact. Once your press release goes live, be prepared to respond to journalists' requests at the drop of a hat. If you do not prepare at least psychologically for a potential flood of interview requests, you might be surprised about how much time it will all take. If and when journalists contact you – either via e-mail or telephone – make sure you respond immediately. Often even 30 minutes is too long before they move onto other stories and you miss your chance. If you are travelling, make sure you have an emergency-contact auto-responder on your e-mail and telephone answer services designed specifically for deadline-enslaved journalists. Usually the initial contact does not include the interview itself – most often the journalist will organise a time and a specific telephone number[8] on which they can call you to pre-record an interview, or schedule it for a live show. Sometimes you will be invited to travel to the nearest studio to be recorded or broadcasted professionally.

Once you do manage to organise a time for the interview, whether it is live radio, recorded television, or just as a chat for a newspaper article, avoid jargon like the plague. In preparation, you can test your language on a non-expert – what is jargon to a non-specialist might not appear to be jargon at all to you. This often comes with experience, but at the very least try to avoid big, technical-sounding words (they do not make you sound more intelligent; rather, they make you sound boring and up-yourself). Still on the issue of language, use short, punchy answers, analogies, and a little humour. Try to relax

---

[7] Nearly all mainstream media organisations call their interviewees thus.

[8] I do not recommend giving out your home or private mobile telephone number to any old journalist; try to keep it to the office if you can. If you do have to do the interview from home and your private line is the only one available, do not indicate that it is a home or private number. Simply tell the journalist that you can be reached at . . .

(again, this comes with experience) by remembering that you know your subject better than nearly everyone who will eventually view or listen to your interview. Also practice the 'Three-Floors Lift Rule' that I wrote about in Chapter 19: if you cannot explain what you have discovered, why this discovery is important, and what aspect of life it might change, within the time it takes a lift to move between three floors, you will not do well in an interview. Remember the structure of your press release and stick to that formula.

Avoid the temptation to take the intellectual higher ground. In other words, there are no stupid questions. You will often be asked the inanest things about your subject;[9] if you are, do not wince, shake your head, or raise an eyebrow. Just politely answer the question clearly, and move on. I cannot stress enough that you will be pitching your story to the deeply uninitiated – i.e., to people who will simultaneously have no experience, understanding, or even the necessary education even to begin to pose relevant questions. It is your challenge to make what you have spent your life studying simple, digestible, and engaging. If the interviewer is struggling to ask relevant, important questions, perhaps suggest a few that he could ask. If it is not a live interview, you will often find that the journalist is happy and relieved to receive your advice.

Sometimes the opposite occurs and a clever (or malevolent) interviewer will ask a question to which you honestly do not know the answer. Resist the temptation to add something if you really do not know, and certainly *never* make anything up! An interviewer with an agenda, or one who has done her homework, can attempt to lead you away from the garden path (i.e., the specific story of your research) by posing questions that go way beyond the results of your study. Of course, you may prognosticate or hypothesise as you see fit, but make sure it is clear that you are not stating any facts if you are not extremely confident of being correct. It is acceptable to

---

[9] A 'journalist' once asked me whether penguins were 'fish or something else'. I had to swallow my guffaw, suppress a condescending smirk, and simply reply 'No. They are birds'. True story.

claim ignorance of a phenomenon or implication by simply saying 'I do not know' during an interview. Stating something as definitive, but eventually proved to be wrong, will come back to haunt you in some unpleasant ways, and potentially tarnish your reputation.

Elaborating on my comment above about 'malevolent' journalists, I want to make it abundantly clear that most journalists are not. However, there is an increasing number of, shall we say, 'skewed' news outlets with particular political agendas at the heart of their reporting philosophies. These sources could not give a tinker's cuss about the validity or accuracy of your results if they can be somehow twisted to support their particular world view. And I am not necessarily restricting my warning to the so-called right-wing outlets; many far-left media sources are also guilty of pursuing less-than-truthful objectives. A clever, but ethically bankrupt journalist will endeavour to make you feel comfortable enough to let down your guard. No matter how friendly, supportive, or understanding this type of journalist might appear to be at face value, always remember that the story (and a twisted, sensationalised one at that) is what he might be after. Such people are not employed to be your friend or do you a service, no matter how disarming their expressions or how deferentially they might address you. If the interviewer tries to hijack the discussion towards a topic that makes you uncomfortable, then it is a simple matter of sticking to your main message and not taking her bait.

If you have agreed to be interviewed by a tabloid or politically motivated source, then expect to be manipulated. A good (tabloid) reporter is highly experienced at trying to make you say things that you will regret for the sake of his employer's ideology. So, understanding as much of your intended audience and who underwrites the news source with whom you are speaking is essential homework for any scientist about to be interviewed by an unfamiliar source. This also means that despite all assurances, absolutely *nothing* is 'off the record'. Even if a journalist says it is 'off record', and she swears up and down that your private comments will remain private, always assume that they will come into the public domain regardless. This warning also applies to even your most trustworthy news sources. However, it is not necessarily a good idea to be highly suspicious of all journalists; merely be aware that there is often an angle, and that journalists are looking for a story that will get them noticed by upper management.

No matter how prepared or wary you might be, in the long run you will inevitably be misquoted, misrepresented, or just plain misunderstood. It happens, so do not let it get to you. I have winced more than once watching or listening to myself in an interview after it has already aired, kicking myself in the process for that stupid thing I said, or that opaque and defensive response I gave. Then I realised that few people apart from me really cared much about what was said, especially in this day of the so-called '24-hour news cycle', after which the once highly important and shocking revelation fades to a vague memory in most people's mind. In other words, do not sweat your mistakes too much.

There can be ways to minimise any damage that might arise once the interview is over. If it is not a live interview, or one that involves recorded voice, then it is always a good idea to ask to see a transcript prior to publication if possible. I estimate that the request works only about half of the time (for the other half, the journalist generally says something like: 'Sorry, we do not do that'). If, however, the journalist is good and wants to write something factual, she will often appreciate a quick check and edit prior to submission to her editor. Regardless, at the very least ask for the internet link to the online version (or podcast if it concerns an interview) so that you can have a record of the event for posterity. You can then use lists of your media appearances as proof of your social engagement – something nearly all institutions ask of their scientists these days.

REGISTER ONLINE

I have written about *active* forms of media engagement almost exclusively in this chapter, because these are by far the most effective means of getting your research noticed by both the public and the greater science community. There are, however, more passive methods to increase your public profile, one of which is through registration at 'expert' media sites.

Most countries today have some form of both free and paid-subscription 'expert' lists that journalists can scan to find people to comment on any subject that comes up. Even most universities provide expert search functions on their websites so that local journalists can find local experts. There are many of these sorts of websites on which a scientist can register as an 'expert' in a particular field, and then journalists can search the database to find someone to talk about the issue of the day. Most of these websites require a paid subscription (prnewswire.com/profnet, expertclick.com, authoratory .com), although there are some free tiers amongst many of them (e.g., expertisefinder.com, helpareporter.com). Of course, there is also Google Scholar on which you should most definitely set up a profile anyway. A savvy journalist could easily find local experts in any field using this free service, but I suspect most of them do not know it even exists.

A little research on the internet will soon give you an idea what is available in your field and region. I like to use scimex.org and expertguide.com.au in Australia, and I know too that many discipline-specific scientific societies offer similar services (some are free; some require payment). Regardless, plastering your name around the internet as a willing and able expert in your field will inevitably lead journalists to ask you to provide comment on the hot news items of the day. In my experience, requests are mostly positive and will provide some public recognition, rather than throw you into a mire of ugly political debate. Of course, you can always refuse requests to comment should they fall at inopportune times or if the story in question is too far outside your particular area of expertise. Just play it by ear.

PARADOX OF MORE = LESS

There is no question at all that science communication has never before been so widespread and of such high quality. More scientists and science students are now blogging, tweeting, and generally engaging the world about their science findings than ever before. There is also an increasing number of professional science communication associations out there, and a growing population of professional science communicators. It is possibly the best time in history to be involved in the generation and communication of scientific results.

Why then is the public appreciation, acceptance, and understanding of science declining? It really does not make much sense if you merely consider that there has never been more good science 'out there' in the media – both social and traditional kinds. For the source literature itself, there has never before been as many scientific journals, articles, and even scientists writing.

So, it is not declining access to information that is the problem, nor is it a retracting body of human scientific knowledge (obviously). Many have pondered why nonsensical political extremism, religiosity, declining educational standards, scientific denialism, conspiracy theories, and evidence-free dogmas are rising despite the public's unparalleled access to knowledge. Few answers are forthcoming or convincing, although it is my own hypothesis that we have finally entered a phase of compensatory resource competition (human density feedback) where the fight to dominate dwindling resources engenders more evidence-free ideologies. In fact, I wager that the first phase of such density feedback in humans is manifested as an ideological response, rather than a behavioural or demographic one (of course, the latter will inevitably follow). So, it is entirely plausible (although still hypothetical at this stage) that despite our increasing frequency and quality of science communication, society is still slipping into an 'Endarkenment'.[10] This potentially negates everything I have ever written about the value of science communication, and certainly would make all those already in the business (or contemplating moving into it) question their life ambitions.

The reality is thankfully a little more encouraging than the simple story told by this gloomy picture. The reason there is a silver lining is that at least from the perspective of actively communicating scientists, the people who matter most in making society-changing policies are more likely to know about their expertise and their good, evidence-based suggestions for improvement than if the scientists only published in academic journals, for hardly anyone outside of academia reads them.

A local example suffices to illustrate this. My country (Australia) has an appalling environmental record and pitiful climate change-mitigation policies (110). Many of our governing politicians are either outright climate-change and/or science denialists, or they could not give a rat's filthy bum about the disastrous state of our planetary life-support system. Yet despite these people and their political and financial agendas, every single municipal and regional council, state government, and even relevant federal ministry has active climate change-mitigation and environmental policies. Of course, many of these policies are entirely inadequate, but at least they exist, and the bureaucrats responsible for their implementation are getting on

---

[10] i.e., the opposite of the 'Enlightenment'.

with their jobs despite the inane politics happening above their heads. When these policy makers need advice and data, they inevitably turn to scientists to provide it. Those scientists who do the best job of communicating their findings are therefore more likely to be noticed by the bureaucrats and invited to contribute their solutions.

These are the people to whom science communicators are now speaking, and thankfully, these are the people that matter most when it comes to making important societal decisions. I therefore urge you not to take science denialism and ideological opposition as reasons for not reaching out to the public with what you do. Instead, make sure that your message is heard clearly by the decision makers who have the most incentive, power, and desire to change society for the better.

PART V What It All Means

23 'Useful' Science

This is where the path to *effective* science becomes a little over-grown and more difficult to follow, especially early in one's career. In Chapter 1, I briefly discussed the concept of 'useful' or 'applied' science, making it clear that I am not insinuating that science must be 'applied' in the classic sense to be considered either useful or important. This sentiment encapsulates the complexity of this issue, and quickly becomes a philosophical endeavour the more one attempts to dissect it. Regardless, most scientists inevitably end up asking themselves this question at some point during their career: 'Is my science useful?' (or more likely, 'What the hell am I doing this for?'). I have indeed asked myself this very question, and I think that I have brokered some sort of philosophical settlement with myself about my (scientific) role on this planet.

To be sure, I was less concerned about the philosophy of it all and more about just doing science when I was a younger man. I certainly did not spend much time contemplating whether the science I was doing was 'useful' or even valued in the eyes of anyone else, because for me it was a personal quest driven by curiosity, competitiveness, a pursuit of intellectualism, and dare I say, and over-inflated sense of self-importance. If other people – be they scientists outside of my particular field or lay people – did not value the scientific questions I was attempting to answer, I was unconcerned; they were clearly deluded Luddites with no appreciation of my finer intellectual pursuits.[1] But for many of the same reasons outlined in Chapter 21, I argue that science – or at least the summarised version of the results it produces – belongs to the people who funded it in the first place,

---

[1] I hope that you detected my sarcastic tone here.

and to a society in general that makes it possible to have scientists research the questions that still elude human understanding.

As I stated in Chapter 1, even the most theoretical research can have immensely important applications to everyday life, even though it might not necessarily be obvious at the time. However, what we generally refer to as 'blue skies' research – scientific questions without obvious applications – are possibly becoming increasingly difficult to fund, despite calls to hold some grant monies aside for just such research (111–113). My experience with granting agencies in more than a dozen countries to date demonstrates without exception that they require an unambiguous description of how your proposed research could benefit society (Chapter 12), even if the direct links to any policies are tenuous. While there are many ways to work the system to your advantage by at least giving the impression that your research will improve the lives of some,[2] my sincere recommendation is that you make the 'application' more than just a superficial coverage, and not just because it might improve your chances of being funded.

The following plea will expose my particular biases, but I think it is worth laying down some arguments about why I think doing at least some *applied* science during your research career will benefit you. My first point is probably the most important of all: while scientists have made impressive leaps in improving the lives of human beings around the world – including *inter alia* reducing mortality and suffering from disease, improving the safety of our technology, making us live more comfortably, and finding better ways to power our societies – we have simultaneously and severely compromised our planet's ability to sustain us comfortably in perpetuity. The massive pressure of 7.5 billion consumers (i.e., people) on our resources and its implications for political stability and our life-support systems in general (114), our uncertain energy future (115, 116), global biodiversity loss (117, 118), mounting pollution and planetary toxification

---

[2] I am not restricting my definition to human beings.

(119), not to mention runaway climate disruption (120), are all painful reminders that we need more than we have ever in the past, big, complex, scientific solutions to these big, complex problems. Remaining pigheadedly closed to the prospect of contributing solutions, even in small ways, to these escalating global problems is therefore both morally and ethically questionable.

Even if you can mount a convincing argument that your own research is applied and that you are contributing to generating solutions, I also urge you to ask yourself to what point you are merely refining well-understood phenomena and avoiding tackling some of the stickier problems facing humanity and our planet. The problem here is that because our problems are generally so scientifically[3] complex to address, it is entirely understandable why many scientists avoid researching them. I think that most of us – as do most people – become comfortable with what they know, and therefore spend most of their time refining their area of expertise. Instead, my sincere opinion is that more of us should jump out of our comfort zones and learn from disciplines outside of our particular, restrictive fields. With even a little more effort here, I think scientists would be far more relevant and successful in turning some of the threatening unsustainability tide back towards more acceptable outcomes. Complicated issues usually take many other disciplines and experts to tackle simultaneously to identify meaningful solutions. I would not expect that a single PhD thesis could do any of these topics real justice, but embedded within a larger framework of multidisciplinary[4] approaches, even the finest-scale research can take on global relevance.

I will not pretend to claim that I have abided by my own advice for much of my career – it is mainly a realisation that has come to me after building a research foundation in possibly more mundane and peripheral research questions. That said, I do not think it merely represents my own maturation and the benefit of hindsight and some

---

[3] Admittedly, the biggest problems are not just scientific in nature, but also include cultural, sociological, economic, and historical elements.

[4] And, *transdisciplinary* (see Chapter 15).

degree of previous success – the issues that we need to dissect and solve are becoming more and more serious every year. My advice is therefore perhaps directed more to the newer generations of scientists; I strongly contend that it will be essential that you take a practical and big-picture approach to your studies and career path over the next several decades. It will take some rather lateral, out-of-the-box thinking on your part to get to grips with these issues, and to engineer solutions that are meaningful at large (including global) scales. Scientists can no longer afford merely to refine our documentation of the planet's demise.

Perhaps the next main reason why you should consider adding an element of practicality to your research is that society is becoming increasingly hostile to intellectuals in general, and to scientists in particular. If you have the good fortune of being a medical researcher examining questions of direct and unambiguous relevance to human well-being, then you might not necessarily appreciate the rise of public backlash against many other scientific disciplines. Although even here there are public outcries against researchers using animal proxies, or heated 'debates' relating to vaccination or surgical intervention. However, the epitome of this anti-science sentiment is public reaction to scientists who have something to do with climate-change research, for no other scientific endeavour today is as politically and ideologically charged, and simultaneously fundamental to the future well-being of people and the lives of every other species on the planet. But it is not just climate science bearing the load of discontent – research as varied as agronomy, evolutionary biology, genetics, astronomy, particle physics, and renewable-energy technology have been the target of (mainly conservative) public derision (121–123). The rising tide of right-wing populist governments worldwide (124) has therefore led to a plethora of challenges to current funding models for science (125, 126), with dissenters cherry-picking particular projects that, to them anyway, appear to be 'wasting' public monies on frivolous, egg-headed pursuits (126).

So clearly, researching something that has an inherent appeal to the broader lay public in terms of practicality might save you a little heartache, although even the most applied research can come under attack if it becomes politicised enough. I return though to my early qualification that there is a strong and justified rationale for doing as much blue-skies research as we can; one must be merely aware of the potential for backlash if your focus is exclusively within the theoretical realm. But *what* you research is only part of the equation regarding public acceptance and a feeling of self-worth and purpose. *How* you go about getting your scientific messages to those people who need it most – what we generically refer to as 'policy makers' – is another matter entirely.

GETTING THE POWERFUL TO LISTEN TO YOU

Even the most applied research, with clear benefits to human society, is more often than not ignored entirely by the very people who need the information the most. If you are more of the theoretical type, than linking your research to real societal change is even more challenging. Part of this communication breakdown arises because of scientists' general lack of training and focus on distilling our frequently complex messages into attractive, bit-sized, digestible chunks for the distracted public's palate. As I described in detail in Part IV, there are many ways you can work to improve your outreach by becoming a better communicator, either through blogging, social media, good public-speaking skills, media prowess, or all of the above. Going above and beyond the peer-reviewed article as the principal medium for your research communication is, of course, essential – but if you leave it there and refuse to blog, be on social media, or become a regular feature in the traditional media, I am afraid your message will not get heard by the people who call the shots. Politicians in particular, and society in general, tend to respect 'experts' with credentials (e.g., lauded track records, lists of prizes, acknowledgements from expert societies, etc.). It might seem superficial to you, I know, but use such accolades to your advantage in the policy realm.

But outreach often goes only so far on the road towards meaningful 'uptake' by the powerful in society – our mayors, regional councillors, government policy-makers, and ultimately, the politicians who govern us. In the pursuit of 'meaning' during your scientific career – if that concerns you at all – one must often go beyond the mildly interesting news byte towards the active engagement of people who make our societies function. Whether you are a medical researcher, a robotics designer, a hydrologist, or a physiologist, I can guarantee that you will eventually become frustrated that the immense effort you have spent to obtain your precious results counts for nought in terms of tangibly solving a real problem. Many applied scientists in particular claim, especially in the first and last paragraphs of their scientific manuscripts and research proposals, that by collecting such-and-such data and doing such-and-such analyses, they will transform how society does business. Unfortunately, most of these claims are hollow or simply untrue, because the results are either (a) never read by the people who actually make decisions, (b) not understood by them, or (c) never implemented because they are too vague or impractical to translate into tangible shifts in policy. And I sincerely hope that once you realise this disappointing state of affairs you will manage your mental health appropriately (see Chapter 18 for some tips on how).

I am not revealing anything shockingly novel here, for we scientists have been discussing the divide between policy-makers and researchers for certainly much longer than I have been alive (127–132), I suspect that we will also be arguing about it long after I am dead too. The main impasse can be summarised succinctly as follows:

SCIENTIST: "Policy makers should get off their arse and read the great papers we publish".

POLICY MAKERS: "Scientists have no idea how policies are made or how to research issues that have real relevance. Their offhand 'advice' is just too naïve."

While there is probably an element of truth to both of these extreme statements, they provide little direction on how to bridge the divide. If you take the time to read the reams of papers and books proffering advice about fixing the problem, then one particular theme emerges: scientists need to do a better job of brokering personal relationships with the people making decisions, because providing mere evidence – no matter how unassailable – neglects how policies are actually formed. If you follow my advice about communicating your science to non-scientists better (Part IV), then you might see why the scientist's lament is also a little bit rich.

But who amongst you has the time or the patience to build personal 'relationships' with policy-makers across the governance gradient? Unless you work for a rather wealthy research institution that employs a team of power brokers to do the rounds on your behalf amongst public servants and politicians in your capital city, you are probably shit-out-of-luck in that regard. So, how can you ever hope to have a chance at making a difference? I am by no means able to claim that much of my work has clearly shifted relevant policies in the right direction, and I will bet that not many scientists can say differently (there are, of course, many great exceptions). But instead of the extreme action of giving up now, there are fortunately some tried and true methods that scientists can employ to swing the policy-uptake pendulum in their favour.

Engagement – that highly abused term referring loosely to *communication with those who will make the policy changes* – is, of course, best done before the research actually starts. Few people like to be told what to do, or worse, be given advice by someone who knows bugger all about the process. How many times have you seen this in a paper? – ' . . . our results demonstrate that policy makers should . . . '. It behoves us then at least to talk to policy people *before* we design our experiments and research questions, and collect our data. It follows then that direct involvement of policy people in the research team from the outset makes particularly good sense. Nothing 'engages' more than when someone has invested time and effort in the research process. The more someone like a public servant invests precious cerebral capacity and time into a project, the greater the chances of them championing the results in their daily mandate. For example, if you were able to talk to even a junior policy maker in your area about what critical information is missing to make her job easier, then it just might shift how you formulate and test particular hypotheses, or how you design your experiments.

But how do you get the opportunity to talk in the first place? One way that requires a bit of foresight and organisation is to set up scoping workshops to which you invite the relevant public servants to participate. These can be as simple as defining the boundaries of the problem needing to be solved, explaining what the various complexities entail, and describing both the scientific and policy-uptake impediments likely restricting real problem solving. Generally speaking, if you can invite policy people with even a modicum of scientific training, your subsequent discussions will tend to be more in the same language, and so consensus will happen more quickly. I would also caution trying to invite someone like an elected official or even their senior aides to the table, because even if they had the time, their understanding of the actual process of day-to-day policy making is probably less developed than a public servant farther down the government food chain. Once you have identified the right people to invite, and they agree, then I recommend making their time spent with you as pleasant as possible; as you recall, I have already described how the ideal workshop can be run (Chapter 3).

It is certainly unnecessary to begin every one of your research projects like this, for you would probably end up wasting too much of your valuable time. Instead, save this sort of 'engagement' for the most policy-tangible research questions for which you have both sufficient money and time to support. Yet another way to involve public servants with some link to policy formation is to consider inviting them to join your lab temporarily, either in the form of an official 'secondment', or even as a student.

A secondment (the temporary transfer of the public servant to your lab to work on particular project under your supervision) is undeniably the easier of the last two options to organise, and it is one that I have used myself to good effect. Of course, this sort of arrangement requires rather a lot of goodwill on the part of the government agency employing the public servant in question, but I have found that in the right circumstances, there is generally a strong desire to 'train' the employee in the relevant scientific methodology. If you can offer your 'services' to an agency in the form of guided supervision of one of their employees, then you will not only engage someone who understands how policy is formulated, you will also have an immensely powerful insider in your camp. There are many additional benefits to such an arrangement that go well beyond the immediate goal of science-policy engagement. First, such people usually bring with them a wealth of data and understanding of the particular problem, and they can greatly enhance the day-to-day culture of your lab members. If you play your cards right from the outset, you can also typically bank on at least a few more research papers coming out of your lab if you coach the person in question to publish their results in the peer-reviewed literature. Another emergent advantage is that a successful secondment will please the government agency to such a point that future research opportunities, and possibly even funding, are more likely to come your way.

In some circumstances, you can even convince the interested public servant to take on an entire degree under your supervision, which of course usually mandates a leave of absence from her place of employment. This approach simply extends the time you have to co-develop the project to mutual satisfaction of both the agency and your lab, but it does come at the cost of making sure the student ends up achieving the degree. I have also had much success adopting this very model, but I have discovered that many such students – most of whom fit within the 'mature' category – never end up going back to their original positions. While this can be a great win for science, it might not please the relevant administration of the government agency. Just make sure you are clear from the outset what your needs and expectations are, and that everyone involved understands the potential risks, issues surrounding intellectual property, and how the eventual results will be used and disseminated.

The other side of this exchange coin is that you yourself can decide to 'second' to work in the government department of interest. I have known more than a few scientists who have done just this, taking anything from six months to several years of leave to work amongst their policy peers in the latter's office building. Most often these kinds of arrangement will require a substantial amount of salary negotiation, but if the will exists on both sides of the fence, it is certainly possible. My main caveat here is that such arrangements can have the opposite effect to 'stealing' public servants to the sciences; there is a substantial risk that you can go missing from the sciences entirely. Not that this is in any way a lesser calling, but be aware that if you stretch yourself too far towards the policy world, you could end up no longer doing the science that got you there in the first place. Also note that a policy organisation is not restricted to a government agency per se, you could also exchange people (including yourself) with non-government organisations, policy think-tanks, or lobby organisations.

It is unlikely in the extreme that as a PhD student, or even a well-published postdoctoral fellow, you will get the attention of a policy-maker directly, unless of course you happen to be working on some highly relevant and topical research project that is the issue of the week in Parliament. However, as you build your reputation within the science community, both in terms of foundational research, but also in how you engage with the public, you just might find that opportunities to become an 'expert' occur more and more frequently. But being recognised as an 'expert' outside of your peer community is far from guaranteed, so do not expect to receive a telephone call from the President or Prime Minister of your country to testify at a Congressional or Parliamentary Hearing. Typically, only the most well-known and highly respected scientists receive this honour,[5] and it might only be a once-in-a-lifetime occasion. If you do find yourself

---

[5] Knowing a few high-profile scientists who have done exactly this, the feelings are rather mixed on whether this constitutes an 'honour'. Being grilled by an agenda-pushing politician with a policy axe to grind is not what most people call 'fun'.

in such a position, the possibilities to influence an entire country's policies are likely higher than any other form of government engagement. One way to increase the probability that you will be noticed in this way is if you regularly make submissions to government agencies when they propose new policies or legislation. If you know more than a bit about the proposal, and if your research can positively inform the outcome, then making even a brief submission with the associated evidence can get the attention of the powerful.

On even rarer occasions, it might be politically expedient to align yourself with the current government's opposition party as a way of standing up to what you deem to be bad policies. A clandestine message or two to a shadow minister about how your scientific results expose the weaknesses of government policy can be like the smell of blood to a vampire. An opposition party will do almost anything to embarrass their political rivals, so tread carefully here – you will make enemies and will most likely be used as political fodder by those who have agendas other than yours. Too close of a partisan alignment can also ostracise you completely from further policy influence once government power shifts. So, while politics should be best left to the politicians, a clever political wielding of your science can have remarkably potent effects, something that I will discuss in much more detail in Chapter 24.

24 Evidence-Based Advocacy

Despite a lot rather uninformed people out there who might view scientists as just flesh-covered automatons lacking the customary set of feelings, we are in fact normal[1] human beings embedded in the same society as everyone else. We own houses, drive cars, have sex, have children, eat in restaurants, drink, dance, vote, pay taxes and utilities, do sport, take vacations, see the doctor, laugh, cry, love, and all the rest of it. As such, we have just as much stake in society as everyone else, and we are very much at the mercy of government policies, cultural norms, and other limitations that everyday life presents us. If we happen to discover through our research some aspect of our society that can be approved, does it suffice merely to publish the material within the academic literature and nebulously 'hope' someone else does some good with it?

If you picked up anything at all from Chapter 23, and understand the benefits of reaching beyond your immediate academic community (Chapters 21 and 22), then it is a short, metaphysical step to entertain the idea of *advocating* for change more proactively than a random tweet or a newspaper interview might imply. I appreciate that the 'a' word strikes fear and derision in the hearts of many scientists, for I too was once under the impression that it was not my job to advocate for anything beyond good scientific practice. Indeed, I practically insisted that my role was uniquely to develop the tools, collect the data, design the experiments, and elaborate the intricacies and complexities of the results to test hypotheses. No more. No less.

---

[1] Who perhaps have a slightly higher incidence of mild personality disorders (Chapter 13).

Even the thought of mimicking those placard-holding street protestors (my rather naïve impression of what an advocate looked like) used to make me figuratively sick to my stomach, for I had formed the opinion that any scientist who took up the protestor's mantle had clearly abandoned his claim to be an intellectual. In my view, once one crossed that intangible line into advocacy, the objectivity of all previous intellectual pursuits was immediately compromised, if not abandoned entirely. For me then, 'advocacy' equalled 'subjectivity', and that was not consistent with what I understood science to be.

I have since done a 180-degree turn on that innocent and narrow-minded perspective, which I imagine does not surprise you having already read up to here. First, I have since come to realise that true objectivity is beyond the reach of any human being, and that science can only provide the tools to reduce our innate subjectivity. As human beings, even scientists have all of our species' weaknesses and limitations of perception, but science allows us to get as close to objectivity as is humanly possible. So, science is not the pursuit of objectivity per se; rather, it is the pursuit of subjectivity reduction. *Errare humanum est.* Whether or not you are willing to admit it, your experiences, culture, and even your language all subtly modify your perception, so given exactly the same question, different scientists will not only try to solve it in different ways, they will also likely come up with slightly different answers. The best we can therefore hope to achieve is to approach some sort of general consensus on how the physical universe operates as all of our different ways of testing approximately converge. Wildly divergent answers to the same question therefore suggest that our biases at least temporarily limit our capacity to understand the 'truth'.

Having established that even you are susceptible to the biases of perception, it might take a little bit of your intellectual elitism off the top. Irrespective of this admission, the next argument for advocacy represents the opposite sentiment – that as a scientist you are, in fact, an 'expert'. In other words, for the topic you research the most, it is likely that there are few others who understand the subtleties, complexities, and intricacies of the problems as well as you do. This places a certain onus on you, the expert, not only to identify where the problems lie, but also what to do about them. Politically ideologies notwithstanding, the scientist is therefore ideally placed as the *best* person to advocate for the solution to a particularly hairy problem. Let us examine a hypothetical example[2] to illustrate. Suppose a toxicologist is testing the effects of various industrial chemicals on the development, health, and survival of some aquatic animal, such as a frog. During the course of her investigations, she discovers that a chemical previously dubbed 'harmless' by its manufacturer, actually compromises the physiology and health of the frog species in question. These effects are known only to her, and perhaps to a small community of other toxicologists once she publishes her results in the peer-reviewed literature. But if that is where she left it, it is unlikely that the negative results would fall into the hands of an environmental advocacy group, even if they managed to decode the statistics and complicated scientific jargon. It is even more unlikely that a regulatory government agency would act on the information to limit the distribution of the chemical, and it is preposterous in the extreme to contemplate that the manufacturer itself would pull the product from the shelves after perusing the latest toxicology literature.

However, if that scientist decided to take a stand against the manufacture and distribution of the harmful chemical by going public with the results and pushing for someone (i.e., government) to do something about it, she would then have a much better chance of seeing positive change compared just to resting on her intellectual laurels. Interviews, submissions to regulatory agencies, opinion editorials, expert testimonials, and other grassroots pushes to publicise the problem, and to advocate for change, are not only within the remit of the scientist, they are arguably a moral obligation for the people best positioned to alert society to the problem. Just as we have a moral obligation to publicise our research to the people who pay for it (i.e.,

---

[2] Although, this example is based on the real experiences of many scientists.

in most cases, the taxpayer), we also have a moral obligation to notify them when we discover that something is not quite right in the world.

Now for the big one – that your objectivity as a scientist is somehow compromised or weakened by your decision to take a stand. But let me be crystal clear here – *how* you do your science, and *what* you do with the results, are two completely different things. This does not, of course, excuse behaviour like painting yourself into an illogical corner by refusing to follow the dominant evidence – some scientists have been caught out here by going so far down the advocacy line that they become bound to their ideology just to save face, rather than follow what the science actually says. While such people may have started to advocate based on the best-available information at the time, the absolute bottom line of the scientist-advocate is that the best-available evidence must be wielded at all times. If you forget, deny, or ignore that simple tenet, then you essentially cease to be a scientist and your advocacy takes on an entirely ideological purpose.

On the other hand, if you do your science to the best of your ability, and you back each of your advocated positions with the dominant scientific evidence, then there is absolutely no chance that your advocacy will compromise the quality or integrity of your work. I am not suggesting that your ideologies will play no role, for the simple choice to advocate itself has an ideological (or at least, moral) foundation. I return to my central position that separating the *how* and the *what* components is the holy grail of the successful and effective scientist-advocate, and that these are in fact rather easy to keep apart as long as they are constantly queried within your conscious forethought. The only real dilemma remaining then is which instruments and actions allow the scientist-advocate to achieve his aims in the most effective manner.

There is considerable overlap between the general communication of your sciences to the masses (Chapter 23) and advocating a position while you do it. For this reason, many of the same tools are also at the scientist-advocate's disposal, such as the 'armchair' advocacy[3] of complaining about something publicly using blogging or social media, or by embedding an advocated position within your regular press engagements. Likewise, submissions to public enquiries can be couched within an advocated position, as long as you accompany it with the best-available scientific evidence. Some more advanced forms of scientific advocacy that require a little more commitment include avenues such as open letters; while still a form of armchair advocacy, such letters can have remarkable power to sway not only the opinions of the public, but also of like-minded advocating agencies and even politicians to do something about the problem in question. Usually letters such as these are position statements – signed by a group of notorious scientists with the relevant know-how – that summarise a particular problem, the scientific evidence on why it is

---

[3] A somewhat derogatory term to describe activities aiming to raise social consciousness from the comfort of one's own home, largely through e-mail and social-media forums. It implicitly contrasts such actions against the more 'active' and traditional forms of street protests.

a problem, and some sort of demand of what should be done about it. The letters are 'open' in the sense that they are not necessarily sent to a single organisation or individual, but they are instead typically published online somewhere, or reproduced in the mainstream media. Open letters of this type have been written on everything from promoting science in society (133) to climate-change mitigation policies (134), and while most probably do no more than solidify existing ideological positions, some of them have had profound implications for changing public opinion and how governments do business.

Slightly more demanding in terms of your time commitments, joining forces with advocacy groups, such as non-government organisations or special-interest societies, can provide you with a veritable army of advocacy expertise and critical mass. At the same time, your scientific expertise can lend credibility and integrity to the group. Exploiting the people-power of the large memberships such groups often enjoy, as well as capitalising on their political contacts and lobbying experience, can launch your evidence-based position to orbits of influence that as an individual you would likely never be able to reach alone. Of course, it is even more important in such relationships to remain vigilant about not compromising your scientific integrity to ideological pressures, and instead insist on following the evidence above all other aims.

On the other hand, effective, science-based advocacy does not necessarily mean that you have to go on the offensive. In fact, some of the most influential advocates are the scientists who work directly with government agencies to change policies 'from within'. Rather than banging on the bureaucrat's door or holding a placard screaming for a politician's head, offering to assist policy-makers by providing data, analytical capacity, or just an informed opinion, can be surprisingly lucrative. Of course, an open-minded and forward-thinking government agency should be inviting scientists to do this sort of thing all the time, but given the political climate in many countries today, it is much more likely that you will be required to be more active and solicit the opportunities yourself. A little savvy positioning and meeting the right people will help, which might be beyond most early-career scientists; however, it is definitely worth keeping in mind that operating collaboratively is sometimes the better option.

Of course, there is nothing stopping the high-integrity scientist-advocate from participating in more traditional forms of activism that involve actions like street protests or sit-ins. I know many of my own colleagues who have done exactly this, and some of them have even been arrested. While that might be taking your average scientist too far out of her comfort zone, it can again be remarkably effective to demonstrate how important a particular issue is to the scientist studying the issue. I would of course advise against throwing Molotov cocktails[4] and looting, but taking part in civil disobedience protests (135), making conscientious objections to war (136), or blocking a railway against coal-carrying boxcars (137) do make decidedly stronger points than signing a petition or complaining online in a single, abbreviated sentence.[5]

I hold the strong opinion that many of civilisation's ailments today are at least partially the result of so few scientists taking a more prominent role in the societies in which they live; remaining passive and passionless observers in the false expectation of preserving objectivity has allowed too many economically and ideologically driven exploiters to take control. Our privilege, knowledge, and education mean that we are morally obliged to assist our fellow human beings by taking, translating, and applying our work to the needs of society as a whole. Without intellectually based solutions, we run the risk that suboptimal and unjust policies will dominate. I am personally not willing to stand idly by, and I encourage all scientists to consider doing likewise where the opportunities present themselves.

[4] Petrol bomb.
[5] Twitter

25 Trials, Tribulations, and Triumphs

After the long wade through the murky swamp of extracurricular requirements to become an effective scientist, you might be asking yourself with renewed vigour why you should even bother. Even the most brilliant scientific mind could easily go unrecognised, unappreciated, and unrewarded by disregarding all or even some of the guidelines I have attempted to outline. Not only must you master the cumulated knowledge, tools, and techniques of your chosen discipline, you must also become a poet, a grammarian, a mathematician, a journalist, a marketer, an entrepreneur, a financier, a socialite, a diplomat, a politician, a protestor, and a leader, while simultaneously maintaining your family relationships and friendships. Being a scientist means that you have to be exceedingly versatile and multi-talented.

Even if you are or can become all of these things, the rewards of a career in academic science are trifling, and at times downright insulting. Universities and many other research organisations are notoriously badly run, flipping uncomfortably and with frustrating frequency between exquisite incompetence and overbearing corporatisation. Even if they were once scientists themselves, your administrators and managers will fail catastrophically to provide you with clear guidance regarding their capricious expectations. You will be underpaid. You will work too much. You will have to fight for every scrap of recognition and freedom. The whopping majority of the students you teach will never even thank you for your efforts. You will also spend your life begging for money to do your research, and in these days of tenuous employment security, you will most likely spend much of your time practically begging to renew your own salary. If your chosen scientific discipline has even a modicum of direct application, you will nearly always be frustrated by the lack of engagement with and recognition by business, politics, and society in general. I used to tell myself that I should never be surprised by the stupidity of my fellow human beings, yet I am surprised almost daily.

Not only will you be largely overlooked, you will more than likely be attacked by those who happen to disagree (ideologically) with your data. I can attest first hand to receiving many threats to my life, my family,[1] and my personal well-being by countless trolls who 'disagree' with my evidence-based recommendations with respect to climate change and environmental management. I have even created a special 'hate mail' folder in my e-mail account for these kinds of unsolicited opinions of my work. Trolls beware – I am keeping them all for posterity. As a result, frustration and even depression are not uncommon states of being for many scientists who choose to engage (as they should) with the general public.

If I could tap into your thoughts right now I would wager an important part of my own anatomy that you are not feeling exactly encouraged to become or remain a scientist. So, I offer you this thought before you throw in the proverbial towel. Despite the bullshit of the daily grind, there is nothing quite as comforting as being aware that science is the only human endeavour that regularly attempts to reduce subjectivity. In the face of all posturing, manipulation, deceit, ulterior motives, and fanatical beliefs that go on every day around us, science remains the bedrock of society, and so despite most human beings being ignorant of its importance, or actively pursuing its demise, all human beings have benefitted from science.[2] What can be more beautiful than finding out how something complex actually works? What can be more assuring than understanding

---

[1] Despite my recommendations to plaster your professional life all over the internet, in this age of Endarkenment and mistrust of scientists, I highly advise that you keep your personal details, such as your home address, the names of your family members, and even the exact location of your office offline.

[2] I have never really liked referring to science as a 'thing' – this falsely gives it a state of entity; rather, science is a 'way' of doing something more objectively than would otherwise be possible. In this sense, it is more a verb than a noun.

that every mystery we encounter always has a logical explanation? It certainly gives me great comfort that there is a way to decipher the complexities of the universe without having to invoke some elaborate nonsense vomited from an overactive imagination. One could confuse such 'comfort' as a type of piety, but unlike religion, science requires updating one's point of view based on evidence, whereas religious faith endures *despite* evidence.

But it is not just the philosophical, or indeed, any sense of superiority that keeps me going. The academic scientist's life can in fact be rather a lot of fun, and it certainly can allow you to see many parts of the world that most others only dream of visiting. While I am not the world's most well-travelled person, my job as a scientist has brought me to many different regions and countries, including Antarctica, Brazil, Canada, Chile, China, Germany, Guatemala, Finland, France, Indonesia, India, Italy, Japan, New Caledonia, New Zealand, Norway, Singapore, South Africa, Spain, Sweden, Tibet, United Kingdom, USA, and all around Australia. And I suspect too that I will see many more countries before my science gig is up. Although I do not do much field work these days, collecting data in some of the most remote and beautiful places of the world has made me appreciate my chosen career path on more than one occasion.

Despite feeling overlooked most of the time, science can also give you a real sense of pride because you know you are doing something useful. Being invited to give seminars to people in almost every walk of life and across age groups – from five years to 95 years – always fills me with at least a little satisfaction and pride. I also appreciate the relative freedom of my profession, despite the excessive number of hours, the grind of grant applications and teaching duties, and the increasing burden of administration. Still, compared to many other careers, being a scientist is comparatively rather flexible, and it rarely feels like I am only working to put food on the table.

Most importantly, the quest to expand the sphere of human knowledge is a joy that I am sure most scientists cite as the primary reason for choosing their career in the first place. We are naturally inquisitive types, so contributing to our own sense of curiosity can be wonderfully fulfilling. If you have even a slightly applied element to your research, you can also take satisfaction in that at least you have justified your existence by contributing something useful to the rest of humankind. While most of us will never be lucky enough to solve some of society's deepest and most intractable problems, we can hold our hand on our hearts and say, 'at least I tried'.

So, whenever I am feeling overly exasperated by the self-serving plutocrats we elect to public office, or when some celebrity spouts evidence-free rubbish about a new miracle cure, or when a religious fanatic lets loose a diatribe of nonsense or a hail of bullets, I try to remember that science (eventually) cuts through all the bullshit. As a scientist, take pride in the knowledge that what you do is like no other human endeavour, and in that knowledge, you can persist in what you do despite society's lack of recognition. If, however, you do go on to win a Nobel Prize for your work, I will claim only the smallest of credit. Thanks for reading, and I earnestly wish you to be happy, successful, and most of all, effective in your chosen life as a scientist.

References

1. Bialystok E, Craik FIM, Luk G (2012) Bilingualism: consequences for mind and brain. *Trends in Cognitive Sciences* 16:240–50.

2. Gold BT, Kim C, Johnson NF, Kryscio RJ, Smith CD (2013) Lifelong bilingualism maintains neural efficiency for cognitive control in aging. *The Journal of Neuroscience* 33:387–96.

3. Laurance WF, Carolina Useche D, Laurance SG, Bradshaw CJA (2013) Predicting publication success for biologists. *BioScience* 63:817–23.

4. Zinsser W (2006) *On Writing Well.* 7th Edition (Harper Collins, New York, NY).

5. Strunk Jr. W, White EB (1999) *The Elements of Style.* (Longman, London).

6. Herrando-Pérez S, Delean S, Brook BW, Bradshaw CJA (2012) Density dependence: an ecological Tower of Babel. *Oecologia* 170:585–603.

7. Herrando-Pérez S, Brook BW, Bradshaw CJA (2014) Ecology needs a convention of nomenclature. *BioScience* 64:311–21.

8. Nuzzo R (2014) Scientific method: statistical errors. *Nature* 506:150–52.

9. Mogie M (2004) In support of null hypothesis significance testing. *Proceedings of the Royal Society of London. Series B-Biological Sciences* 271:S82–S84.

10. Elliott LP, Brook BW (2007) Revisiting Chamberlain: multiple working hypotheses for the 21st century. *BioScience* 57:608–14.

11. Burnham KP, Anderson DR (2002) *Model Selection and Multimodel Inference: A Practical Information-Theoretic Approach.* 2nd Edition (Springer-Verlag, New York, NY).

12. Burnham KP, Anderson DR (2004) Understanding AIC and BIC in model selection. *Sociological Methods and Research* 33:261–304.

13. Link WA, Barker RJ (2006) Model weights and the foundations of multimodel inference. *Ecology* 87:2626–35.

14. Lukacs PM, Thompson WL, Kendall WL, et al. (2007) Concerns regarding a call for pluralism of information theory and hypothesis testing. *Journal of Applied Ecology* 44:456–60.

15. Gaertner-Johnston L (2006) "That" or "Which"? *Business Writing.* www.businesswritingblog.com/business_writing/2006/01/that_or_which.html (accessed 17 October 2017).

16. Tang M (2011) *Passive voice vs. active voice.* sciencewritingblog.wordpress
    .com/2011/04/12/passive-voice-vs-active-voice (accessed 17 October 2017).

17. Day RA, Gastel B (2012) *How to Write and Publish a Scientific Paper.* 7th
    Edition (Cambridge University Press, Cambridge).

18. Hofmann AH (2009) *Scientific Writing and Communication: Papers, Propos-
    als, and Presentations.* 1st Edition (Oxford University Press, Oxford).

19. Schimel J (2012) *Writing Science. How to Write Papers that Get Cited and
    Proposals that Get Funded.* (Oxford University Press, Oxford).

20. Bradshaw CJA, Brook BW (2016) How to rank journals. *PLoS One*
    11:e0149852.

21. Pautasso M (2013) Ten simple rules for writing a literature review. *PLoS Com-
    putational Biology* 9:e1003149.

22. Vetter D, Rücker G, Storch I (2013) Meta-analysis: a need for well-defined
    usage in ecology and conservation biology. *Ecosphere* 4:1–24.

23. Laurance WF, Useche DC, Rendeiro J, et al. (2012) Averting biodiversity col-
    lapse in tropical forest protected areas. *Nature* 489:290–4.

24. Atlas Collaboration, CMS Collaboration, Aad G, et al. (2015) Combined mea-
    surement of the Higgs Boson mass in pp collisions at $\sqrt{s} = 7$ and 8 TeV with
    the ATLAS and CMS experiments. *Physical Review Letters* 114:191803.

25. Ball P (2016) The mathematics of science's broken reward system. *Nature*
    doi:10.1038/nature.2016.20987.

26. Ware M, Mabe M (2015) The STM report: an overview of scientific and schol-
    arly journal publishing. Copyright, Fair Use, Scholarly Communication, etc.
    (International Association of Scientific, Technical and Medical Publishers,
    The Hague, Netherlands).

27. Jinha AE (2010) Article 50 million: an estimate of the number of scholarly
    articles in existence. *Learned Publishing* 23:258–63.

28. Landhuis E (2016) Scientific literature: information overload. *Nature*
    535:457–8.

29. Pyke GH (2013) Struggling scientists: cite our papers! *Current Science*
    105:1061–6.

30. Pyke GH (2014) Achieving research excellence and citation success: what's
    the point and how do you do it? *BioScience* 64:90–1.

31. PLoS Medicine Editors (2006) The Impact Factor game. *PLoS Medicine*
    3:e291.

32. Seglen PO (1997) Why the impact factor of journals should not be used for
    evaluating research. *British Medical Journal* 314:497.

33. Jacsó P (2008) The pros and cons of computing the h-index using Web of Sci-
    ence. *Online Information Review* 32:673–88.

34. Ramírez A, García E, Del Río J (2000) Renormalized Impact Factor. *Scientometrics* 47:3–9.

35. Althouse BM, West JD, Bergstrom CT, Bergstrom T (2009) Differences in impact factor across fields and over time. *Journal of the American Society for Information Science and Technology* 60:27–34.

36. Neff BD, Olden JD (2010) Not so fast: inflation in Impact Factors contributes to apparent improvements in journal quality. *BioScience* 60:455–9.

37. Bergstrom CT, West JD, Wiseman MA (2008) The Eigenfactor^{TM} Metrics. *The Journal of Neuroscience* 28:11433–4.

38. Moed HF (2010) Measuring contextual citation impact of scientific journals. *Journal of Informetrics* 4:265–77.

39. González-Pereira B, Guerrero-Bote VP, Moya-Anegón F (2010) A new approach to the metric of journals' scientific prestige: the SJR indicator. *Journal of Informetrics* 4:379–91.

40. Guerrero-Bote VP, Moya-Anegón F (2012) A further step forward in measuring journals' scientific prestige: the SJR2 indicator. *Journal of Informetrics* 6:674–88.

41. Hirsch JE (2005) An index to quantify an individual's scientific research output. *Proceedings of the National Academy of Sciences of the USA* 102:16569–72.

42. Braun T, Glänzel W, Schubert A (2006) A Hirsch-type index for journals. *Scientometrics* 69:169–73.

43. Delgado-López-Cózar E, Cabezas-Clavijo Á (2013) Ranking journals: could Google Scholar Metrics be an alternative to Journal Citation Reports and Scimago Journal Rank? *Learned Publishing* 26:101–13.

44. Falagas ME, Kouranos VD, Arencibia-Jorge R, Karageorgopoulos DE (2008) Comparison of SCImago journal rank indicator with journal impact factor. *The FASEB Journal* 22:2623–8.

45. Jacsó P (2008) The pros and cons of computing the h-index using Google Scholar. *Online Information Review* 32:437–52.

46. Jacsó P (2008) The pros and cons of computing the h-index using Scopus. *Online Information Review* 32:524–35.

47. Aksnes DW, Sivertsen G (2004) The effect of highly cited papers on national citation indicators. *Scientometrics* 59:213–24.

48. Aksnes DW (2003) Characteristics of highly cited papers. *Research Evaluation* 12:159–70.

49. Sumner P, Vivian-Griffiths P, Boivin, J, et al. (2016) Exaggerations and caveats in press releases and health-related science news. *PLoS One* 11:e0168217.

50. Osterloh M, Kieser A (2015) Double-blind peer review: how to slaughter a sacred cow. *Incentives and Performance: Governance of Research*

Organizations, eds Welpe MI, Wollersheim J, Ringelhan S, and Osterloh M (Springer International Publishing, Cham), pp 307–21.

51. Faggion Jr CM (2016) Improving the peer-review process from the perspective of an author and reviewer. *British Dental Journal* 220:167–8.

52. Didham RK, Leather SR, Basset Y (2017) Don't be a zero-sum reviewer. *Insect Conservation and Diversity* 10:104.

53. Schmitt J (2015) Can't disrupt this: Elsevier and the 25.2 billion dollar a year academic publishing business. *Medium.com*. medium.com/@jasonschmitt/can-t-disrupt-this-elsevier-and-the-25-2-billion-dollar-a-year-academic-publishing-business-aa3b9618d40a#.kbdywg8ns (accessed 17 October 2017).

54. Relx Group (2016) Annual Reports and Financial Statements 2015. (mslgroup.co.uk, London, United Kingdom) www.relx.com/investorcentre/reports%202007/Documents/2015/relxgroup_ar_2015.pdf (accessed 17 October 2017).

55. Larivière V, Haustein S, Mongeon P (2015) The oligopoly of academic publishers in the digital era. *PLoS One* 10:e0127502.

56. John Wiley and Sons Inc. (2016) John Wiley and Sons, Inc. and Subsidiaries Form 10-K for the Fiscal Year Ended April 30, 2016. (Wiley-Blackwell, Hoboken, New Jersey, USA) wiley.com/WileyCDA/Section/id-370237.html (accessed 17 October 2017).

57. Springer (2013) Springer Science+Business Media. General Overview and Financial Performance 2012. (Springer, Berlin, Germany) static.springer.com/sgw/documents/1412702/application/pdf/Annual_Report_2012_01.pdf (accessed 17 October 2017).

58. Sage (2016) Sage Group plc Preliminary Results for the year ending 30 September 2015. (Sage, Newcastle Upon Tyne, United Kingdom) www.sage.com/#/media/group/files/Sage%20Group%20Results%202015.pdf (accessed 17 October 2017).

59. Holcombe A (2015) Scholarly publisher profit update. *Alex Holcombe's Blog.* alexholcombe.wordpress.com/2015/05/21/scholarly-publisher-profit-update (accessed 17 October 2017).

60. de Vries J (2012) Thousands of scientists vow to boycott Elsevier to protest journal prices. *ScienceInsider* www.sciencemag.org/news/2012/02/thousands-scientists-vow-boycott-elsevier-protest-journal-prices (accessed 17 October 2017).

61. Van Noorden R (2014) The scientists who get credit for peer review. *Nature* doi:10.1038/nature.2014.16102.

62. Wilkinson MD, Dumontier M, Aalbersberg IJ, et al. (2016) The FAIR Guiding Principles for scientific data management and stewardship. *Scientific Data* 3:160018.

63. Hollowell J, Nicholas G (2008) Intellectual property issues in archaeological publication: some questions to consider. *Archaeologies* 4:208–17.

64. Lindenmayer D, Scheele B (2017) Do not publish. *Science* 356:800.

65. Teferra D, Altbachl PG (2004) African higher education: challenges for the 21st century. *Higher Education* 47:21.

66. Bhandari R, Blumenthal P eds (2011) *International Students and Global Mobility in Higher Education: National Trends and New Directions.* (Palgrave MacMillan, New York, NY).

67. Slippers B, Vogel C, Fioramonti L (2015) Global trends and opportunities for development of African research universities. *South African Journal of Science* 111:a0093.

68. Altbach PG (2013) Brain drain or brain exchange? *International Higher Education* 72:2–4.

69. Grace OM (2017) Crowdfunding your science. *Nature Ecology and Evolution Community* natureecoevocommunity.nature.com/users/54273-olwen-m-grace/posts/18355-crowdfunding-your-science (accessed 17 October 2017).

70. Fitzpatrick SM, Bruer JT (1997) Science funding and private philanthropy. *Science* 277:621.

71. Weingart P, Guenther L (2016) Science communication and the issue of trust. *Journal of Science Communication* 15:1–11.

72. Cook I, Grange S, Eyre-Walker A (2015) Research groups: how big should they be? *PeerJ* 3:e989.

73. Conti A, Liu CC (2015) Bringing the lab back in: personnel composition and scientific output at the MIT Department of Biology. *Research Policy* 44:1633–44.

74. Zakaib GD (2011) Science gender gap probed. *Nature* 470:153.

75. Ceci SJ, Williams WM (2011) Understanding current causes of women's underrepresentation in science. *Proceedings of the National Academy of Sciences of the USA* 108:3157–62.

76. Helmer M, Schottdorf M, Neef A, Battaglia D (2017) Gender bias in scholarly peer review. *eLife* 6:e21718.

77. Knobloch-Westerwick S, Glynn CJ, Huge M (2013) The Matilda Effect in science communication. *Science Communication* 35:603–25.

78. Moss-Racusin CA, Dovidio JF, Brescoll VL, Graham MJ, Handelsman J (2012) Science faculty's subtle gender biases favor male students. *Proceedings of the National Academy of Sciences of the USA* 109:16474–9.

79. Aldercotte A, Guyan K, Lawson J, Neave S, Altorjai S (2017) ASSET 2016: experiences of gender equality in STEMM academia and their intersections

with ethnicity, sexual orientation, disability and age. (Equality Challenge Unit, London, United Kingdom).

80. Grunspan DZ, Eddy SL, Brownell SE, Wiggins BL, Crowe AJ, Goodreau SM (2016) Males under-estimate academic performance of their female peers in undergraduate biology classrooms. *PLoS One* 11:e0148405.

81. Kimmel MS (2009) Gender equality: not for women only. *Equality, Diversity and Inclusion at Work: A Research Companion*, ed Èzbilgin MF (Edward Elgar Publishing, Cheltenham, United Kingdom), pp 359–71.

82. Holter ØG (2013) Masculinities, gender equality and violence. *Masculinities and Social Change* 2:51–81.

83. Miller T, del Carmen Triana M (2009) Demographic diversity in the board-room: mediators of the board diversity-firm performance relationship. *Journal of Management Studies* 46:755–86.

84. McGuire KL, Primack RB, Losos EC (2012) Dramatic improvements and persistent challenges for women ecologists. *BioScience* 62:189–96.

85. O'Brien KR, Hapgood KP (2012) The academic jungle: ecosystem modelling reveals why women are driven out of research. *Oikos* 121:999–1004.

86. Mayer AL, Tikka PM (2008) Family-friendly policies and gender bias in academia. *Journal of Higher Education Policy and Management* 30:363–74.

87. Martin JL (2014) Ten simple rules to achieve conference speaker gender balance. *PLoS Computational Biology* 10:e1003903.

88. Favaro B, Oester S, Cigliano JA, et al. (2016) Your science conference should have a code of conduct. *Frontiers in Marine Science* 3:103.

89. Davis KS (1999) Why science? women scientists and their pathways along the road less traveled. *Journal of Women and Minorities in Science and Engineering* 5:129–53.

90. Handelsman J, Cantor N, Carnes M, et al. (2005) More women in science. *Science* 309:1190.

91. Benderly B (2011) A Nobel Laureate's advice to women scientists. *Science Magazine Blog*. blogs.sciencemag.org/sciencecareers/2011/06/its-a-good-thin.html (accessed 17 October 2017).

92. Folkins CH, Sime WE (1981) Physical fitness training and mental health. *American Psychologist* 36:373–89.

93. Hillman CH, Erickson KI, Kramer AF (2008) Be smart, exercise your heart: exercise effects on brain and cognition. *Nature Reviews Neuroscience* 9:58–65.

94. Deslandes A, Moraes H, Ferreira C, et al. (2009) Exercise and mental health: many reasons to move. *Neuropsychobiology* 59:191–8.

95. Kramer AF, Erickson KI (2007) Capitalizing on cortical plasticity: influence of physical activity on cognition and brain function. *Trends in Cognitive Sciences* 11:342–8.

96. Brisswalter J, Collardeau M, René A (2002) Effects of acute physical exercise characteristics on cognitive performance. *Sports Medicine* 32:555–66.

97. Kashihara K, Maruyama T, Murota M, Nakahara Y (2009) Positive effects of acute and moderate physical exercise on cognitive function. *Journal of Physiological Anthropology* 28:155–64.

98. Hillman CH, Motl RW, Pontifex MB, et al. (2006) Physical activity and cognitive function in a cross-section of younger and older community-dwelling individuals. *Health Psychology* 25:678–87.

99. Penedo FJ, Dahn JR (2005) Exercise and well-being: a review of mental and physical health benefits associated with physical activity. *Current Opinion in Psychiatry* 18:189–93.

100. Fontaine KR (2000) Physical activity improves mental health. *The Physician and Sportsmedicine* 28:83–4.

101. Saxena S, Van Ommeren M, Tang KC, Armstrong TP (2005) Mental health benefits of physical activity. *Journal of Mental Health* 14:445–51.

102. Callaghan P (2004) Exercise: a neglected intervention in mental health care? *Journal of Psychiatric and Mental Health Nursing* 11:476–83.

103. Weyerer S, Kupfer B (1994) Physical exercise and psychological health. *Sports Medicine* 17:108–16.

104. Dudo A, Besley JC (2016) Scientists' prioritization of communication objectives for public engagement. *PLoS One* 11:e0148867.

105. Anonymous (2010) Closing the Climategate. *Nature* 468:345.

106. Eysenbach G (2011) Can tweets predict citations? Metrics of social impact based on Twitter and correlation with traditional metrics of scientific impact. *Journal of Medical Internet Research* 13:e123.

107. Shuai X, Pepe A, Bollen J (2012) How the scientific community reacts to newly submitted preprints: article downloads, Twitter mentions, and citations. *PLoS One* 7:e47523.

108. Priem J, Costello KL (2010) How and why scholars cite on Twitter. *Proceedings of the American Society for Information Science and Technology* 47:1–4.

109. Thelwall M, Haustein S, Larivière V, Sugimoto CR (2013) Do altmetrics work? Twitter and ten other social web services. *PLoS One* 8:e64841.

110. Bradshaw CJA, Ehrlich PR (2015) *Killing the Koala and Poisoning the Prairie: Australia, America, and the Environment.* (University of Chicago Press, Chicago).

111. Linden B (2008) Basic blue skies research in the UK: are we losing out? *Journal of Biomedical Discovery and Collaboration* 3:3.

112. House of Lords (2010) Setting priorities for publicly funded research. Volume II: Evidence. (United Kingdom House of Lords, The Stationery Office Limited, London, United Kingdom).

113. Bhattacharya A (2012) Science funding: duel to the death. *Nature News* 488:20–2.

114. Bradshaw CJA, Brook BW (2014) Human population reduction is not a quick fix for environmental problems. *Proceedings of the National Academy of Sciences of the USA* 111:16610–15.

115. Martínez DM, Ebenhack BW (2008) Understanding the role of energy consumption in human development through the use of saturation phenomena. *Energy Policy* 36:1430–5.

116. Heard BP, Brook BW, Wigley TML, Bradshaw CJA (2017) Burden of proof: a comprehensive review of the feasibility of 100% renewable-electricity systems. *Renewable and Sustainable Energy Reviews* 76:1122–33.

117. WWF (2016) Living Planet Report 2016. (WWF, Gland, Switzerland) wwf .panda.org/about_our_earth/all_publications/lpr_2016 (accessed 17 October 2017).

118. Pimm SL, et al. (2014) The biodiversity of species and their rates of extinction, distribution, and protection. *Science* 344:1246752.

119. Cribb J (2014) *Poisoned Planet*. (Allen & Unwin, Crows Nest, New South Wales, Australia).

120. IPCC (2014) Climate Change 2014: Impacts, Adaptation, and Vulnerability. IPCC WGII AR5 Technical Summary. (Intergovernmental Panel on Climate Change, Geneva, Switzerland).

121. Lehmann E (2012) Conservatives lose faith in science over last 40 years. *Scientific American* www.scientificamerican.com/article/conservatives-lose-faith-in-science-over-last-40-years (accessed 17 October 2017).

122. Dierkes M, von Grote C eds (2000) *Between Understanding and Trust: The Public, Science and Technology*. (Routledge, London, United Kingdom).

123. Master Z, Resnik DB (2013) Hype and public trust in science. *Science and Engineering Ethics* 19:321–35.

124. Oliver JE, Rahn WM (2016) Rise of the Trumpenvolk. *The ANNALS of the American Academy of Political and Social Science* 667:189–206.

125. Lamberts R, Grant WJ (2011) The government's war on science: deliberate attack, or abuse by neglect? *The Conversation* theconversation.com/the-governments-war-on-science-deliberate-attack-or-abuse-by-neglect-208 (accessed 17 October 2017).

126. Mervis J (2016) Senator's attack on 'cheerleading' study obscures government's role in training scientists. *ScienceInsider* doi:10.1126/science.aaf9993.

127. Cairney P, Oliver K (2016) If scientists want to influence policymaking, they need to understand it. *The Guardian* www.theguardian.com/science/political-science/2016/apr/27/if-scientists-want-to-influence-policymaking-they-need-to-understand-it (accessed 17 October 2017).

128. Shanley P, López C (2009) Out of the loop: why research rarely reaches policy makers and the public and what can be done. *Biotropica* 41:535–44.

129. Gibbons P, Zammit C, Youngentob K, et al. (2008) Some practical suggestions for improving engagement between researchers and policy-makers in natural resource management. *Ecological Management and Restoration* 9:182–6.

130. Garvin T (2001) Analytical paradigms: the epistemological distances between scientists, policy makers, and the public. *Risk Analysis* 21:443–56.

131. van den Hove S (2007) A rationale for science–policy interfaces. *Futures* 39:807–26.

132. Choi BCK, Pang T, Lin V, et al. (2005) Can scientists and policy makers work together? *Journal of Epidemiology and Community Health* 59:632.

133. Science and Technology Australia (2017) Open Letter for Science. (Science and Technology Australia) scienceandtechnologyaustralia.org.au/open-letter-for-science (accessed 17 October 2017).

134. Caldeira K, Emanuel K, Hansen J, Wigley T (2013) Top climate change scientists' letter to policy influencers. *CNN* edition.cnn.com/2013/11/03/world/nuclear-energy-climate-change-scientists-letter/index.html (accessed 17 October 2017).

135. Bryner J (2013) NASA climate scientist arrested in pipeline protest. *Live Science* www.livescience.com/27117-nasa-climate-scientist-arrest.html (accessed 17 October 2017).

136. Maynard-Casely H (2017) Inspiring to speak out – two physicists who changed the world. *The Conversation* theconversation.com/inspiring-to-speak-out-two-physicists-who-changed-the-world-72654 (accessed 17 October 2017).

137. Frid A (2012) Conservation value of paddy wagon currency: civil disobedience by scientists. *Conservation Bytes* conservationbytes.com/2012/05/12/conservation-value-of-paddy-wagon-currency (accessed 17 October 2017).

Index